Choosing and Using S
A Biologist's Guide

Calvin Dytham

Department of Biology, University of York

b

Blackwell
Science

© 1999 by
Blackwell Science Ltd
Editorial Offices:
Osney Mead, Oxford OX2 0EL
25 John Street, London WC1N 2BL
23 Ainslie Place, Edinburgh EH3 6AJ
350 Main Street, Malden
   MA 02148 5018, USA
54 University Street, Carlton
   Victoria 3053, Australia
10, rue Casimir Delavigne
   75006 Paris, France

Other Editorial Offices:
Blackwell Wissenschafts-Verlag GmbH
Kurfürstendamm 57
10707 Berlin, Germany

Blackwell Science KK
MG Kodenmacho Building
7–10 Kodenmacho Nihombashi
Chuo-ku, Tokyo 104, Japan

The right of the Author to be
identified as the Author of this Work
has been asserted in accordance
with the Copyright, Designs and
Patents Act 1988.

First published 1999
Reprinted 2000, 2001

Set by Setrite Typesetters,
Hong Kong
Printed and bound in Great Britain
by MPG Books Ltd, Bodmin

The Blackwell Science logo is a
trade mark of Blackwell Science Ltd,
registered at the United Kingdom
Trade Marks Registry

DISTRIBUTORS

Marston Book Services Ltd
PO Box 269
Abingdon, Oxon OX14 4YN
(*Orders*: Tel: 01235 465500
         Fax: 01235 465555)

USA
Blackwell Science, Inc.
Commerce Place
350 Main Street
Malden, MA 02148 5018
(*Orders*: Tel: 800 759 6102
             781 388 8250
         Fax: 781 388 8255)

Canada
Login Brothers Book Company
324 Saulteaux Crescent
Winnipeg, Manitoba R3J 3T2
(*Orders*: Tel: 204 837-2987)

Australia
Blackwell Science Pty Ltd
54 University Street
Carlton, Victoria 3053
(*Orders*: Tel: 3 9347 0300
         Fax: 3 9347 5001)

A catalogue record for this title
is available from the British Library

ISBN 0-86542-653-8

Library of Congress
Cataloging-in-publication Data

Dytham, Calvin.
   Choosing and using statistics:
   a biologists guide/
   Calvin Dytham.
      p.   cm.
   Includes bibliographical references
   ISBN 0-86542-653-8
   1. Biometry. I. Title.
QH323.5.D98   1998
570′.1′5195 — dc21   98-28514
                                        CIP

For further information on
Blackwell Science, visit our website:
www.blackwell-science.com

Contents

Preface, ix

1 Eight steps to successful data analysis, 1

2 The basics, 2
Observations, 2
Hypothesis testing, 2
P values, 3
Sampling, 3
Experiments, 4
Statistics, 4
   Descriptive, 4
   Tests of difference, 4
   Tests of relationships, 5
   Tests for data investigation, 5

3 Choosing a test: a key, 7
Remember—eight steps to successful data analysis, 7
The art of choosing a test, 7
A key to assist in your choice of statistical test, 7

4 Hypothesis testing, sampling and experimental design, 22
Hypothesis testing, 22
Acceptable errors, 22
P-values, 23
Sampling, 23
   Choice of sample unit, 24
   Number of sample units, 24
   Positioning of sample units to achieve a random sample, 24
   Timing of sampling, 25
Experimental design, 26
   Control, 26
      Procedural controls, 26
      Temporal control, 27
      Experimental control, 27
      Statistical control, 27
   Some standard experimental designs, 28

5 Statistics, variables and distributions, 30
What are statistics?, 30
Types of statistics, 30
What is a variable?, 31
Types of variables or scales of measurement, 31

Measurement variables, 32
  How accurate do I need to be?, 33
Ranked variables, 33
Types of distributions, 34
  Why do you need to know about distributions?, 34
  Discrete distributions, 34
    The Poisson distribution, 34
    The binomial distribution, 34
    Negative binomial distribution, 36
    Hypergeometric distribution, 37
  Continuous distributions, 37
    The rectangular distribution, 37
    The normal distribution, 37
  Transformations, 40
  The *t*-distribution, 40
    Confidence intervals, 41
  The chi-square distribution, 41
  The exponential distribution, 41
  Nonparametric 'distributions', 42
    Ranking, quartiles and the interquartile range, 42

6 Descriptive and presentational techniques, 43
Displaying data: summarizing a single variable, 43
Displaying data: showing the distribution of a single variable, 43
Descriptive statistics, 46
  Statistics of location, 46
  Statistics of distribution, dispersion or spread, 48
  Other summary statistics, 49
Using the computer packages, 50
  General, 50
Displaying data: summarizing two or more variables, 53
Displaying data: comparing two variables, 53
  Associations, 53
  Trends, predictions and time series, 56
Displaying data: comparing more than two variables, 57
  Associations, 57
  Multiple trends, time series and predictions, 59

7 The tests 1: tests to look at differences, 61
Do frequency distributions differ?, 61
  Questions, 61
  G-test, 61
  Chi-square test (χ^2), 61
    An example, 62
  Kolmogorov–Smirnov test, 70
    An example, 70
  Anderson–Darling test, 72
Do the observations from two groups differ?, 72
  Paired data, 72

Paired *t*-test, 73
Wilcoxon's signed ranks test, 75
Unpaired data, 80
t-test, 80
One-way ANOVA, 85
Mann–Whitney *U*, 92
Do the observations from more than two groups differ?, 95
Repeated measures, 95
Friedman test (for repeated measures), 95
Repeated measures ANOVA, 98
Independent samples, 98
One-way ANOVA, 99
Kruskal–Wallis test, 106
Post hoc testing: after one-way ANOVA, 108
There are two independent ways of classifying the data, 110
One observation for each factor combination (no replication), 111
An example, 111
Friedman test, 111
Two-way ANOVA (without replication), 114
ANOVA with more than one observation for each factor combination (with replication), 120
Interaction, 121
Two-way ANOVA (with replication), 123
Scheirer–Ray–Hare test, 131
There are more than two independent ways to classify the data, 136
Multifactorial testing, 137
Three-way ANOVA (without replication), 137
Three-way ANOVA (with replication), 137
Multiway ANOVA, 142
Not all classifications are independent, 142
Nested or hierarchical designs, 143
Two-level nested design ANOVA, 143

8 The tests 2: tests to look at relationships, 147

Is there a correlation or association between two variables?, 147
Observations assigned to categories, 147
Chi-square test of association, 147
Cramér coefficient of association, 153
Phi coefficient of association, 153
Observations assigned a value, 153
'Standard' correlation (Pearson's product–moment correlation), 154
Spearman's rank-order correlation, 158
Kendall's rank-order correlation, 160
Regression, 162
Is there a 'cause and effect' relationship between two variables?, 162
'Standard' linear regression, 163
Kendall's robust line-fit method, 169
Logistic regression, 169
Model II regression, 170

Polynomial and quadratic regression, 170
Tests for more than two variables, 170
  Tests of association, 170
    Correlation, 170
    Partial correlation, 171
    Kendall's partial rank-order correlation, 171
Cause(s) and effect(s), 171
    Regression, 172
    ANCOVA (analysis of covariance), 172
    Multiple regression, 175
    Stepwise regression, 175
    Path analysis, 176

9   The tests 3: tests for data exploration, 177
Types of data, 177
    Observation, inspection and plotting, 177
    Principal component analysis (PCA) and factor analysis, 177
    Canonical variate analysis (CVA), 182
    Discriminant function analysis, 182
    Multivariate analysis of variance (MANOVA), 185
    Multivariate analysis of covariance (MANCOVA), 185
    Cluster analysis, 185
    DECORANA and TWINSPAN, 187

10   Symbols and letters used in statistics, 188
Greek letters, 188
Symbols, 188
Upper case letters, 189
Lower case letters, 190

11   Assumptions of the tests, 191
What if the assumptions are violated?, 194

12   Hints and tips, 195
Using a computer, 195
Sampling, 196
Statistics, 196
Displaying the data, 197

Glossary, 199

Bibliography and short reviews of selected texts, 211

Index, 215

Preface

The purpose of this book

My aim was to produce a statistics book with two characteristics: to assume that the reader is using a computer to analyse data and to contain absolutely no equations.

This book is a handbook for biologists who want to process their data through a statistical package on the computer to select the most appropriate methods and extract the important information from the often confusing output that is produced. It is aimed, primarily, at undergraduates and masters students in the biological sciences who have to use statistics in practical classes and projects. Such users of statistics don't have to understand exactly how the test works or how to do the actual calculations, and these things are not covered in this book as there are more than enough books that do that already. What is important is that the right statistical test is used and the right inferences made from the output of the test. An extensive key to statistical tests is included for the former and the bulk of the book is made up of descriptions of how to carry out the tests to address the latter.

In several years of teaching statistics to biology students it is clear to me that most students don't really care how or why the test works. They do care a great deal that they are using an appropriate test and interpreting the results properly. I think that this is a fair aim to have for occasional users of statistics. Of course, anyone going on to use statistics frequently should become familiar with the way that calculations manipulate the data to produce the output as this will give a better understanding of the test.

If this book has a message it is this: think about the statistics *before* you collect the data! So many times I have seen rather distraught students unable to analyse their precious data because the experimental design they used was inappropriate. On such occasions I try to find a compromise test that will make the best of a bad job but this often leads to a weaker conclusion than might have been possible if more forethought had been applied from the outset. There is no doubt that if experiments or sampling strategies are designed with the statistics in mind better science will result.

Statistics are often seen by students as the 'thing you must do to data at the end'. Please try to avoid falling into this trap yourself. Thought experiments producing dummy data are a good way to try out experimental designs and are much less labour intensive than real ones!

Although there are no equations in this book I'm afraid there was no way to totally avoid statistical jargon. To ease the pain somewhat, an extensive glossary and key to symbols is included. So when you are navigating your way through the key to choosing a test you should look up any words you don't understand.

In this book I have given extensive instructions for the use of three widely available software packages: SPSS, Excel and MINITAB. However, the key to choosing a statistical test is not at all package-specific so if you use a software package other than the three I focus on or if you are using a calculator you will still be able to get a good deal out of this book.

If every sample gave the same result there would be no need for statistics. All aspects of biology seemed to be filled with variation. It is statistics that can be used to penetrate the haze of experimental error and inherent variability to reach the underlying causes and processes at work. So try not to hate statistics — they are merely a tool that, when used wisely and properly can make the life of a biologist much simpler and give conclusions a sound basis.

How to use this book

This is definitely not a book that should be read from cover to cover. It is a book to refer to when you need assistance with statistical analysis, either when choosing an appropriate test or when carrying it out. The basics of statistical analysis and experimental design are covered briefly but those sections are intended mostly as a revision aid or outline of some of the more important concepts. The reviews of other statistics books may help you choose those that are most appropriate for you if you want or need more details.

The heart of the book is the key. The rest of the book hinges on the key, explaining how to carry out the tests, giving assistance with the statistical terms in the glossary or giving tips on the use of computers and packages.

Packages used

MINITAB version 10.51, MINITAB inc.
SPSS versions 6.1 and 7.0, SPSS inc.
Excel version 5.0, Microsoft Corporation
Windows ® versions 3.11 and 95 (release 2), Microsoft Corporation

Example data

In the spirit of dummy data collection, all example data used throughout this book has been fabricated. Any similarity to data alive or dead is purely coincidental.

Acknowledgements

Thanks to Sheena McNamee for support during the writing process, to Andrea Gillmeister and two anonymous reviewers for commenting on an early version of the manuscript and to Terry Crawford, Jo Dunn, David Murrell and Josephine Pithon for recommending and lending various books. Thanks also to Ian Sherman and Susan Sternberg at Blackwell Science and to many of my colleagues who told me that the general idea of a book like this was a sound one. Finally, I would especially like to thank the students at the University of York, UK who brought me the problems that provided the inspiration for this book.

Calvin Dytham
York

1: Eight steps to successful data analysis

This is a very simple sequence that, if you follow it, will integrate the statistics you use into the process of scientific investigation. As I make clear here, statistical tests should be considered *very early* in the process and not left until the end.

1 Decide what you are interested in.

2 Formulate a hypothesis or several hypotheses (see Chapters 2 and 3 for guidance).

3 Design the experiment, manipulation or sampling routine that will allow you to test the hypotheses (see Chapters 2 and 4 for some hints on how to go about this).

4 Collect dummy data (i.e. make up approximate values based on what you expect to obtain). The collection of 'dummy data' may seem strange but it will convert the proposed experimental design or sampling routine into something more tangible. The process can often expose flaws or weaknesses in the data collection routine that will save a huge amount of time and effort.

5 Use the key (presented in Chapter 3) to guide you towards the appropriate test or tests.

6 Carry out the test(s) using the dummy data. (Chapters 6–9 will show you how to input the data, use the statistical packages and interpret the output.)

7 If there are problems go back to step 3 (or 2), otherwise proceed to the collection of the real data.

8 Carry out the test(s) using the real data. Report the findings and/or return to step 2.

I implore you to use this sequence. I have seen countless students who have spent a long time and a lot of effort collecting data only to find that the experimental or sampling design was not quite right. The test they are forced to use is much less powerful than one they could have used with only a slight change in the experimental design. This sort of experience tends to turn people away from statistics and become 'scared' of them. This is a great shame as statistics are a hugely useful and vital tool in science.

The rest of the book follows this eight-step process, but you should use it for guidance and advice when you become unsure of what to do.

2: The basics

The aim of this section is to introduce, in rather broad terms, some of the recurring concepts of data collection and analysis. Everything introduced here is covered at greater length in later chapters and in the many statistics text books that aim to introduce statistical theory and experimental design to scientists.

The key to statistical tests in the next chapter assumes that you are familiar with most of the basic concepts introduced here.

Observations

Observations are the raw material of statistics. These can be anything that is recorded as a part of an investigation. They can be on any scale from a simple 'raining or not raining' dichotomy to a very sophisticated and precise analysis of nutrient concentrations. The type of observations recorded will have a great bearing on the type of statistical tests that are appropriate.

Observations can be simply divided into three types: *categorical* where the observations can be in a limited number of categories which have no obvious scale (e.g. 'oak', 'ash', 'elm'); *discrete* where there is a real scale but not all values are possible (e.g. 'number of eggs in a nest' or 'number of species in a sample') and *continuous* where any value is theoretically possible, only restricted by the measuring device (e.g. lengths, concentrations). Different types of observations are considered in more detail in Chapter 5.

Hypothesis testing

The cornerstone of scientific analysis is hypothesis testing. The concept is rather simple; almost every time a statistical test is carried out it is testing the probability that a hypothesis is correct. If the probability is small then the hypothesis is deemed to be untrue and it is rejected in favour of an alternative. This is done in what seems to be a rather upside down way as the test is always of what is called the null hypothesis rather than the more interesting hypothesis. The null hypothesis is the hypothesis that nothing is going on (it is often labelled as H_0). For example, if the weights of bulbs for two cultivars of daffodils were being investigated, the null hypothesis would be that there is nothing going on: 'the weights of the two groups of bulbs are the same'. A statistical test is carried out to find out how likely that null hypothesis is to be true. If we decide to reject the null hypothesis we must accept the alternative, interesting hypothesis (H_1) that: 'the weights of bulbs for the two cultivars are different'.

P values

The *P* value is the bottom line of most statistical tests. It is simply the probability that the hypothesis being tested is true. So, if a *P* value is given as 0.06 that indicates that the hypothesis has a 6% chance of being true. In biology it is usual to take a value of 0.05, or 5%, as the critical level for the rejection of a hypothesis. This means that providing a hypothesis has a less than one in twenty chance of being true we reject it. As it is the null hypothesis that is nearly always being tested we are always looking for low *P* values to reject this hypothesis and accept the more interesting alternative hypothesis.

Clearly, the smaller the *P* value the more confident we can be in the conclusions drawn from it. A *P* value of 0.0001 indicates that the chance of the hypothesis being tested being true is one in ten thousand, and this is much more convincing than a marginal $P = 0.049$.

P values and the types of errors that are implicitly accepted by their use are considered further in Chapter 4.

Sampling

Observations have to be collected in some way. This process of data acquisition is called sampling. There are almost as many different methods that can be used for sampling as there are possible things to sample although there are some general rules. One of the first is that more observations are better than few, but balanced sampling is even more important (i.e. when comparing two groups take the same number of observations from each group).

Most statistical tests assume that samples are taken at random. This sounds easy but is actually quite difficult to achieve. For example if you are sampling beetles from pitfall traps the sample may seem totally random but in fact is quite biased in favour of those individuals that are moving around the most. Another common bias is to choose a point at random and then sample the nearest individual to that point, assuming that this will produce a random sample. It will not be random at all as individuals at the edges of clumps are more likely to be selected than those in the middle. There are methods available to reduce problems associated with nonrandom sampling but the first step is to be aware of the problem.

A further assumption of sampling is that individuals are either only measured once or they are all sampled on several occasions. This assumption is often violated if, for example, the same site is visited on two occasions and the same individuals or clones are inadvertently remeasured.

The sets of observations collected are called variables. A variable can be almost anything it is possible to record as long as different individuals can be assigned different values.

Some of the problems of sampling are considered in Chapter 4.

Experiments

In biology many investigations use experiments of some sort. An experiment occurs if anything is altered or controlled by the investigator. For example, an investigation into the effect of fertilizer on plant growth will use a control plot (or several control plots) where there is no fertilizer added and then one or several plots where fertilizer has been added at a known concentration set by the investigators. In this way the effect of fertilizer can be determined by comparison of the different concentrations of fertilizer. The condition being controlled (e.g. fertilizer) is usually called a factor and the different levels used called treatments or factor levels (e.g. concentrations of fertilizer). The design of this experiment will be determined by the hypothesis or hypotheses being investigated. If the effect of the fertilizer on a particular plant is of interest, then perhaps a range of different soil types might be used with and without fertilizer. If the effect on plants in general is of interest, then an experiment using a variety of plants is required, perhaps in isolation or together. If the optimum fertilizer treatment is required then a range of concentrations will be applied and a cost–benefit analysis carried out. Details of the experimental design are considered in more detail in Chapter 4.

Statistics

In general, statistics are the results of manipulation of observations to produce a single, or small number of results. There are various categories of statistics depending on the type of summary required.

Descriptive

The simplest type of statistics are summaries of sets of data. Simple summary statistics are easy to understand but should not be overlooked. These are not usually considered to be statistics but are, in fact, extremely useful for data investigation. The most widely used are measures of the 'location' of a set of numbers, such as the mean or median. Then there are measures of the 'spread' of the data, such as the standard deviation. Choice of appropriate descriptive statistic and the best way of displaying the results are important and are considered in Chapters 5 and 6.

Tests of difference

A familiar question in any field of investigation is going to be something like: '*is this group different from that group?*'. A question of this kind can then be turned into a null hypothesis with a form: '*this group and that group are not different*'. To answer this question and test the null hypothesis a statistical test of difference is required. There are many tests that all seem to answer the same type of question, but each is appropriate when certain types of data are being considered. After the simple comparison of two groups there are extensions to comparisons of more than two groups and then to tests involving more than one way of dividing the individuals into groups. For example, individuals could be assigned to two groups by sex and also into groups depending on whether they had been given a drug or not. This could be considered as four groups or as what is known as a factorial test where there are two factors 'sex' and 'drug' with all combinations of the levels of the two factors being measured in some way. Factorial designs can become very complicated, but they are very powerful and can expose subtleties in the way the factors interact that can never be found through simple investigation of the data.

Tests of difference can also be used to compare variables with known distributions. These can be statistical distributions or derived from theory. Tests of difference are considered in more detail in Chapter 7.

Tests of relationships

Another familiar question that arises in scientific investigation is in the form: '*is A associated with B?*'. For example, 'is fat intake related to blood pressure?'. This type of question should then be turned into a null hypothesis that '*A is not associated with B*' and then tested using one of a variety of statistical tests. As with tests of difference there are many tests that seem to address the same type of problem, but again each is appropriate for different types of data.

Tests of relationships fall into two groups called correlation and regression depending on the type of hypothesis being investigated. Correlation is a test to measure the degree to which one set of data varies with another — it *does not* imply that there is any 'cause' and 'effect' relationship. Whereas regression is used to fit a relationship between two variables such that one can be predicted from the other. This *does* imply a 'cause' and 'effect' relationship or at least an implication that one of the variables is a 'response' in some way. So, in the investigation of fat intake and blood pressure a strong positive correlation between the two shows a relationship but does not show cause and effect. If there is a significant positive regression line this would imply that blood pressure can be predicted using fat intake *or* if the regression uses the fat intake as the 'response' that, rather implausibly, that fat intake can be predicted from blood pressure.

There are many additional techniques that can be employed to consider the relationships between more than two sets of data. Tests of relationships are considered in Chapter 8.

Tests for data investigation

A whole range of tests are available to help investigators explore large data sets. Unlike the tests considered above, data investigation need not have a hypthesis for testing. For example, in a study of the morphology of fish there may be many fin measures from a range of species and sites that offer far too many potential hypotheses for investigation. In this case the application of a multivariate technique may show up relationships between individuals, help assign unknown specimens to categories or just suggest which hypotheses are worth further consideration.

A few of the many different techniques available are considered in Chapter 9.

3: Choosing a test: a key

I hope that you are not reading this section with your data already collected and the experiment or sampling programme 'finished'. If you have finished collecting your data I strongly advise you to approach your next experiment or survey in a different way. As you will see below, I hope that you will be using this key *before* you start collecting real data.

Remember—eight steps to successful data analysis

1   Decide what you are interested in.
2   Formulate a hypothesis or hypotheses.
3   Design the experiment, manipulation or sampling routine.
4   Collect dummy data. Make up approximate values based on what you expect.
5   Use the key here to decide on the appropriate test or tests.
6   Carry out the test(s) using the dummy data.
7   If there are problems go back to step 3 (or 2), otherwise collect the real data.
8   Carry out the test(s) using the real data.

The art of choosing a test

It may be a surprising revelation but choosing a test is not an exact science. There is nearly always scope for considerable choice and many decisions will be made based on personal judgements, experience with similar problems or just plain hunches. There are many circumstances where there are several ways that the data could be analysed and yet each of the possible tests could be justified.

A common tendency is to force the data from your experiment into a test you are familiar with even if it is not the best method. Look around for different tests that may be more appropriate to the hypothesis you are testing. In this way you will expand your statistical repertoire and add power to your future experiments.

A key to assist in your choice of statistical test

Starting at point 1 move through the key following the path that best describes your data. If you are unsure about any of the terms used then consult the glossary or the relevant sections of the next two chapters.

There may be several end points appropriate to your data that result from this key. For example you may wish to know the correct display method for the data and then the correct measure of dispersion to use. If this is the case, go through the key twice. All the tests and techniques mentioned in the key are described in later chapters.

Italics indicate instructions about what you should do.

Numbers in brackets indicate that the point in the key is something of a compromise destination.

There are several points where rather arbitrary numbers are used to determine which path you should take. For example, I use 30 different observations as the level at which to split continuous and discontinuous data. If your data set falls close to this level you should not feel constrained to take one path if you feel more comfortable with the other.

---

1   Testing a clear hypothesis and associated null hypothesis (e.g. H_1 = blood glucose level is related to age and H_0 = blood glucose is not related to age).   25

Not testing any hypothesis but simply want to present, summarize or explore data.   2

2   Methods to summarize and display the data required.   3

Data exploration for the purpose of understanding and getting a feel for the data or perhaps to help with formulation of hypotheses. For example, you may wish to find possible groups within the data (e.g. ten morphological variables have been taken from a large number of carabid beetles, the multivariate test may establish whether they can be divided into separate taxa).   60

3   There is only one collected variable under consideration (e.g. the only variable measured is brain volume although it may have been measured from several different populations).   4

There is more than one variable (e.g. you have measured the number of algae/ml *and* the water pH in the same sample).   24

4   The data are discrete; there are less than 30 different values (e.g. number of species in a sample).   5

The data are continuous; there are more than 29 different values (e.g. bee wing length to the nearest 0.01 mm).   16
(Note: the distinction between the above is rather arbitrary.)

5   There is only one group or sample (e.g. all measurements taken from the same river on the same day).   6

There is more than one group or sample (e.g. you have measured the number of antenna segments in a species of beetle and have divided the sample according to sex to give two groups).   15

6   Graphical representation of the data is required.   7

Numerical summary or description of the data required.   11

7   A display of the whole distribution required.   8

Crude display of position and spread of data required: *use a box and whisker display, page 43 (also known as a box plot).*

8 Values have real meaning (e.g. number of mammals caught per night).   10

Values are arbitrary labels that have no real sequence (e.g. different land classifications in an area of forest).   9

9 There are less than 10 different values/classifications: *draw a pie chart, page 45. Ensure that each segment is labelled clearly and that adjacent shading patterns are as distinct as possible. Avoid using 3D (three-dimensional) or shadow effects, dark shading or colour. Do not add the proportion in figures to the 'piece' of the pie as this information is redundant.*

There are 10 or more different values/classifications: *amalgamate values until there are less than 10 or divide the sample to produce two sets each with less than 10 values. Ten is a level above which it is difficult to distinguish different sections of the pie or to have sufficiently distinct shading patterns.*

10 There are more than 20 different values: *group to produce around 12 classes (almost certainly done automatically by your package) and draw a histogram, page 44. Classes on x-axis, frequency of occurrence (number of times the value occurs) on y-axis, no gaps between bars. Do not use 3D or shadow effects.*

There are 20 or fewer different values: *draw a bar chart, page 43. Each value should be represented on the x-axis, if there are few classes extend the range to include values not in the data set at either side, frequency of occurrence (number of times the value occurs) on y-axis, gaps between bars unless the variable is clearly supposed to be continuous, no 3D or shadow effects.*

11 Measure of position (mean is the one used most commonly).   12

Measure of dispersion or spread (standard deviation and confidence intervals are the most commonly used).   13

Measure of symmetry or shape of the distribution.   14

(Note: you will probably want to go for at least one measure of position and another of spread in most cases.)

12 Variable is definitely discrete, usually restricted to integer values less than 30 (e.g. number of eggs in a clutch): *calculate the median, page 47.*

Variable should be continuous but has only a few different values due to accuracy of measurement (e.g. bone length measured to the nearest centimetre): *calculate the mean, page 46.*

If you are particularly interested in the most commonly occurring response: *calculate the mode, page 47, in addition to either the mean or median.*

[9]

13 Very rough measure of spread required: *calculate the range, page 48 (note that this measure is very biased by sample size and is rarely a useful statistic).*

You are particularly interested in the highest and/or lowest values: *calculate the range, page 48.*

Variable should be continuous but has only a few values due to accuracy of measurement: *calculate the standard deviation, page 48.*

Variable is discrete or has an unusual distribution: *calculate the interquartile range, page 48.*

14 Variable should be continuous, but has only a few values because of accuracy of measurement: *calculate the skew* (g_1), *page 49.*

Observations are discrete or if you have already calculated the interquartile range and the median: *the relative size of the interquartile range above and below the median provides a measure of the symmetry of the data.*

15 You have not established the appropriate technique for a single sample: *go back to 6 to find the appropriate techniques for each group. You should find the same is correct for each sample or group.* (6)

The samples can be displayed separately: *go back to 7 and choose the appropriate style. So that direct comparisons can be made, be sure to use the same scales (both x-axis and y-axis) for each graph. Be warned that packages will often adjust scales for you. If this happens you must force the scales to be the same.* (7)

The samples are to be displayed together on the same graph: *use a chart with a box plot for each sample and the x-axis representing the sample number, page 53. Make sure there is a clear space between each box plot.*

16 There is only a data set from one group or sample. 17

The data has been collected from more than one group or sample (e.g. you have measured the mass of each individual of a single species of vole from one sample and have divided the sample according to sex). 23

17 A graphical representation of the data is required. 18

Numerical summary or descriptive statistics are required. 19

18 Display of the whole distribution required: *group to produce around 12–20 classes and draw a histogram, page 44 (probably done automatically by your package). Classes on x-axis, frequency of occurrence (number of times the value occurs within the class) on y-axis, no gaps between bars, no 3D or shadow effects. Even-sized classes are much easier for a reader to interpret. Data with an unusual distribution (e.g. there are some extremely high*

values well away from most of the observations) may require transformation before the histogram is attempted.

A crude display of position and spread of the data is required: the 'error bar' type of display is unusual for a single sample but common for several samples. There is a symbol representing the mean and a vertical line representing range of either the 95% confidence interval or the standard deviation, page 53.

19   Measure of position (mean is the most common).   20
     Measure of dispersion (spread).   21
     Measure of symmetry or shape of the distribution.   22

You wish to determine whether the data is normally distributed. *Carry out a Kolmogorov–Smirnov test, page 70, an Anderson–Darling test, page 72 or a chi-square goodness of fit, page 61.*

(Note: you probably require one of each of the above for a full summary of the data.)

20   Unless the variable is definitely discrete or is known to have an odd distribution (e.g. not symmetrical): *calculate the mean, page 46.*

If the data are known to be discrete or the data set is to be compared with other, discrete data with fewer possible values: *calculate the median, page 47.*

If you are particularly interested in the most commonly occurring value: *calculate the mode, page 47, in addition to the mean or median.*

21   If the data are continuous and approximately normally distributed and you require an estimate of the spread of data: *calculate the standard deviation (SD), page 48.*

(Note: standard deviation is the square root of variance and is measured in the same units as the original data.)

If you have previously calculated the mean and require an estimate of the range of possible values for the mean: *calculate 95% confidence limits for the mean, page 49 (a.k.a. 95% confidence interval or 95% CI).*

Very rough measure of spread required: *calculate the range, page 48. (Note that this measure is very biased by sample size and is rarely a useful statistic in large samples.)*

If you have a special interest in the highest and or lowest values in the sample: *calculate the range, page 48.*

If the data is known to be discrete or is to be compared with other, discrete data or if you have previously calculated the median: *calculate the interquartile range, page 48.*

(Note: many people use standard error (SE) as a measure of spread. I am convinced that the main reason for this is that it is smaller than either SD or 95% CI, rather than for any statistical reason. Do not use standard error for this purpose.)

22 If the data is continuous and approximately normally distributed and you require an unbiased estimate of the symmetry of the data: *calculate the skewness/asymmetry of the data* (g_1), *page 49.*
If the data is continuous and approximately normally distributed, you have calculated skewness and require an estimate of the 'shape' of the distribution of the data: *calculate the kurtosis* (g_2), *page 49. (This is rarely required as a graphical representation will give a better understanding of the shape of the data. Kurtosis is only really worth calculating in samples with more than 100 observations.)*
If you have already calculated the interquartile range and the median: *re-examine the interquartile range. The relative size of the interquartile range above and below the median provides a measure of the symmetry of the data.*

23 You have not established the appropriate technique for a single sample: *go back to 16 to find the appropriate techniques for each of the groups. You should find the same is appropriate for each sample or group.*   (16)
The samples can be displayed separately: *go back to 17 and choose the appropriate style. So that direct comparisons can be made, be sure to use the same scales (both x-axis and y-axis) for each graph. Be warned that packages will often adjust scales for you.*   (17)
The samples are to be displayed together on the same graph: *use a chart with an 'error bar' (showing the mean and a measure of spread) for each sample and the x-axis representing the sample number or site. Do not join the means unless intermediate samples would be possible (i.e. don't join means from samples divided by sex or species but do join those representing temperature or nutrient concentration if the intervals between different sample temperatures are even).*

24 If each variable is to be considered separately: *go back to 4 and consider each variable in turn.*   (4)
Two variables only: *a two-dimensional scatter plot can be drawn, page 54. The choice of variable for x and y-axes is free but if you suspect a possibility of 'cause' and 'effect' the 'cause' should be on the x-axis. Do not draw a line of best fit even if it is offered by the package unless the situation is appropriate and you have carried out a regression analysis.*
Three variables: *a 3D scatter plot can be drawn, page 57. It is very difficult to represent three dimensions on a two-dimensional sheet of paper or computer screen. You must drop spikes to the 'floor' of the graph otherwise it is impossible to visualize the spread in the third dimension. It may be better to use a series of two-dimensional scatters instead.*

More than three variables: *use a series of two- or three-dimensional scatterplots, page 57.*

25 (Note: the distinction here will be slightly fuzzy in some cases, but essentially there are two basic types of test.)

The hypothesis is investigating differences, and the null hypo-thesis that there is no difference between groups or between data and a particular distribution (e.g. H_1 [hypothesis] white-eye and carmine-eye flies have different development times, H_0 [null hypothesis] white-eye and carmine-eye flies have the same development time).   26

The hypothesis is investigating a relationship and the null hypo-thesis is that there is no relationship (e.g. H_1 [hypothesis] plant size is related to available phosphorus in the soil, H_0 [null hypothesis] plant size is not related to amount of phosphorus).   46

26 Data are collected as individual observations (e.g. height in cm).   29
Data are in the form of frequencies (e.g. when carrying out a plant cross and scoring the number of offspring of each flower type).   27

27 There are two possibilities only (e.g. wild type or mutant).   28
There are more than two possibilities: *carry out a G-test if your package supports it, page 61, otherwise use a chi-square goodness of fit, page 61.*
There are more than about five possibilities: *a Kolmogorov–Smirnov test, page 70, may be more convenient than the chi-square goodness of fit, page 61.*

28 There are more than 200 observations in the sample: *carry out a G-test, page 61, if your package supports it, otherwise use a chi-square goodness of fit, page 61.*
There are 25 to 200 observations: *carry out a G-test if your package supports it, page 61; otherwise use a chi-square goodness of fit, page 61, but add a 'continuity correction' by adding 0.5 to the lower frequencies and subtracting 0.5 from the higher. This is very conservative and may result in a non-significant result when a marginally significant one is present (Type I error). If your package supports the 'Williams correction' then use that instead of the 'continuity correction'.*
Less than 25 observations: *four possible solutions in order of preference: use a binomial test if supported by your package; carry one out by hand if you are able; get a bigger sample; pretend you have 25 observations and use the instructions above.*

29 There is only one way of classifying the data (e.g. grouped by species).   30
There is more than one way of classifying the data (e.g. grouped by species and sex).   38

30   There are only two groups (e.g. male and female or 'before' and   31
     'after').
     There are more than two groups (e.g. later samples from four   35
     different ponds).
     (Note: the null hypothesis is that all groups have the same mean
     so if any two groups have different means you have to reject this
     null hypothesis.)
     There are more than two groups and several measured variables
     (e.g. individuals divided by species (a grouping variable) and the
     measured variables are various anatomical characters or dimensions
     such as leaf length, stem thickness and petal length): *canonical
     variate analysis, page 182.*

31   Two samples are 'paired'. This means that the same individual   32
     has been measured twice. This is the 'before' and 'after' design
     (e.g. river nitrate level is measured before and after a storm).
     Two samples are independent. There are different groups of   34
     individuals in the two samples.

32   The data are normally distributed, there are at least 30 possible
     values and variances are, at least approximately, homogeneous:
     carry out a paired t-test, *page 73. To test for normal distribution
     use Kolmogorov–Smirnov test, page 70, an Anderson–Darling
     test, page 72, or a chi-square goodness of fit, page 61. A test for
     homogeneity of variance is often an option within the* t-test *in the
     package (e.g. a Levene test or Bartlett's test).
     A two-way ANOVA test is a potential alternative here but is more
     difficult to carry out than the paired* t-test *in most statistical
     packages, page 98 (use one factor of the ANOVA to represent
     'before/after' and the other to represent the different individuals).*
     Above does not, or might not, apply.   33

33   Data have more than 20 possible values: *carry out a Wilcoxon
     signed ranks test, page 75.*
     Data have less than 20 possible values (e.g. questionnaire results
     with a question of 'how do you feel' asked before and after
     exercise): *carry out a sign test if supported by your package (this
     is a very conservative but fairly low power test), page 77. If this is
     not available in the package carry out a Wilcoxon signed ranks
     test, page 75.*

34   The data set is normally distributed, there are at least 30 possible
     values and variances are, at least approximately, homogeneous:
     *carry out a one-way ANOVA with one factor having two levels
     (one for each group), page 85, or use a* t-test, *page 80. To test for
     normal distribution use Kolmogorov–Smirnov tests, page 70,
     Anderson–Darling tests, page 72, or chi-square goodness of fit,
     page 61. A test for homogeneity of variance is often an option*

within the t-test *or the ANOVA in the package (e.g. a Levene test).
The traditional method is to use a* t-test *for this type of experiment
but it is no better than ANOVA in this circumstance as both tests
give an identical result.*
The data set does not or might not fulfil the conditions above:
*carry out a Mann–Whitney U-test, page 92 (sometimes called
Wilcoxon–Mann–Whitney or Wilcoxon two-sample). The Kruskal–
Wallis test is a possible alternative but is less powerful.*

35 Samples are 'repeated measures'. This means that the same   36
individual is measured through time. This is an extended 'before'
and 'after' design (e.g. lake turbidity is measured each year for
several years).
Each sample is independent. There are different groups of   37
individuals in each samples.
It is important that no individual is present more than once in the
data set, otherwise problems (of inappropriate replication) reduce
the power of the statistical test.

36 The data set is normally distributed, there are at least 30 possible
values and variances are, at least approximately, homogeneous:
*carry out a two-way ANOVA with one factor having a different
level for each sampling repeat and a second factor having a level
for each individual you are sampling, page 85 (easy if you have
only five rivers measured each year but very tedious to input and
difficult to interpret if you have 50). To test for normal distributions
you can use Kolmogorov–Smirnov tests, page 70, Anderson–Darling
tests, page 72, or a chi-square goodness of fit, page 61, although
in practice it is usual to use experience to determine whether the
data is likely to be normally distributed. Furthermore the ANOVA
is quite robust to small departures from a normal distribution.*
The data set does not conform to the restrictions above and you
only have one observation for each repeat of each sample: *carry
out a Friedman test with one factor having a different level for
each sampling repeat event, page 111 (e.g. before, during, after),
and one factor having a different level for each individual (e.g.
person) you are sampling.*
Neither of the above apply. This one is difficult! It results from
poor planning of the experiment: *usually it is best to carry out an
ANOVA, page 123, as if the data conformed to the assumptions of
distribution and variances but to treat the resulting P-values with
caution especially if the P-value is between 0.1 and 0.01.*

37 The data set is normally distributed, there are at least 30 possible
values and variances are, at least approximately, homogeneous:
*carry out a one-way ANOVA with the one factor having one level
for each group, page 99. (Note: the* t-test *can only be used on two*

groups.) If the result is significant then you need to carry out a post-hoc test to determine which factor levels are significantly *different from which, page 108. If you are cautious, or unsure, use a Kruskal–Wallis test instead, page 106.*

The data set does not, or might not, fulfil the conditions above: *carry out a Kruskal–Wallis test with one factor having a level for each group, page 106. (Note: the Mann–Whitney U-test only works for two groups so is not appropriate here.)*

38   There are only two factors/ways of classifying the data (e.g. strain and location).   39

There are three factors/ways of classifying the data (e.g. sex, region and year).   43

There are more than three factors: *use the key as if there are three factors and extrapolate. Multifactorial experimental designs become increasingly difficult to interpret and it is often easiest to leave out factors that you have proved to have no significant effect, page 142.*   (43)

39   There is no replication (i.e. only one value assigned to each combination of the two factor levels, e.g. the basal trunk diameters after 2 years are collected from four strains of apple tree grown under four watering regimes but with only one tree under each watering condition).   40

There is replication (i.e. there are two or more values for each combination of the two factors).   41

40   The data are likely to be normally distributed (it is almost impossible to test this).

Data such as lengths and concentrations are likely to be appropriate but judgement is required: *carry out a two-way ANOVA, page 114 (but note that you will not be able to look for any interaction between the two factors).*

You are cautious or have a data set that is unlikely to be normally distributed: *carry out a Friedman test, page 111.*

41   Factors are fully independent of each other.   42

One factor is 'nested' within another (e.g. if there are three branches sampled from each of three trees then branch is said to be 'nested' within trees): *carry out a nested ANOVA, page 143, (a.k.a. hierarchical ANOVA). (Note there is no nonparametric (i.e. one that makes fewer assumptions about the distribution of the data) equivalent of this test.)*

42   The data set is normally distributed, there are at least 30 possible values and variances are approximately equal: *two-way ANOVA, page 123, measure of the interaction between the two factors is possible.*

The data set is not as above. Versions of a two-way ANOVA making fewer assumptions about the data (i.e. nonparametric) are only a recent innovation and are not yet appearing in statistical packages. If the experiment is balanced, that is there are the same number of observations for each combination of factor levels: *carry out a Scheirer–Ray–Hare test, page 131. This will, almost certainly, not be in your statistical package but can still be carried out with a little modification of other tests. See the section describing the test for details.*

43 (Note: there are no nonparametric tests available from here on so if the data set does not fit the assumptions of the test you have no alternatives. ANOVA is quite robust to failure to meet its assumptions but be aware, especially if results are borderline.)

All factors (ways of grouping the data) are independent of each other.   44

At least one factor is nested within another. (For example, in an experiment the variable is blood sugar level in mice. The factors are litter, female and food provided. If there are two litters from each of two females then litter will be 'nested' within female. Neither litter nor female will be 'nested' within food.)   45

44 There is only one observation for each combination of factor levels: *carry out a three-way ANOVA, page 137. You will not be able to calculate the significance of the three-way interaction but you will be able to do this for the interaction between each combination of two factors. (Note that any main factors that prove to be nonsignificant may be left out of the analysis to reduce the complexity of the design.)*

There is more than one observation for each combination of factor levels: *carry out a three-way ANOVA, interaction terms are possible, page 137.*

45 One factor is 'nested' within another, the third is independent (as in the mouse example in 43): *carry out an ANOVA involving both hierarchical and crossed factors. This is often difficult to reach in statistics packages although the design is a common one. If you only have one observation for each combination of factor levels then an interaction term cannot be tested (this is because it has to be used as the residual or error term).*

One factor is 'nested' within another that is itself 'nested' within a third. (For example, in a water pollution survey the variable is nitrate concentration. Several samples have been taken from five streams from each of three river systems and this has been done in two countries. The factors are 'stream', 'river' and 'country'. 'Stream' nested within 'river' which is nested within 'country'): *carry out a nested or hierarchical ANOVA, page 143.*

46 (The choice you have here is one that is frequently confused, be careful.)

The purpose of the test is to look for an association between variables (e.g. is there an association between wing length and thorax length?). You have not set (controlled) one of the variables in the experiment. There is no reason to assume a 'cause' and 'effect' relationship. This is a correlation type of test.   47

One or more of the variables has been set (controlled or selected) by the experiment or there is a probable 'cause' and 'effect' or functional relationship between variables.   53

One of the uses of regression statistics you are moving to is prediction (e.g. the experiment is looking at the effect of temperature on heart rate in *Daphnia*. You are expecting that heart rate is affected by temperature but wish to discover the form of the relationship so that predictions can be made). This is a regression type of test.

47 Data are in the form of frequencies (e.g. number of 'white flowers' and 'orange flowers').   48

There is a value for each observation. Variables should be paired etc. (e.g. an observation of two variables 'cell count' and 'lung capacity' from one individual).   50

48 There are two variables: *if you follow this thread further you will reach tests that are often awkward to carry out in packages and are often easier to calculate by hand. If you do calculate them by hand you may have to look up the significance level using a chi-squared table.*   49

There are more than two variables: *simultaneous comparisons of frequencies for more than two classifications are very difficult to interpret. It is best to compare them pair-wise.*

49 The two variables each have two possible (e.g. 'yes'/'no' or 'male'/'female'): *calculate a phi coefficient for a 2 × 2 table, page 153, if your package supports it or you can do it by hand. This test is a special case of a contingency chi-square calculation, page 147.* At least one of the variables has more than two possible values (e.g. a crude land classification 'forest'/'scrub'/'pasture'/'arable' is compared to an estimate of the density of a small mammal 'common'/'rare'/'absent'): *calculate a contingency chi-square, page 147, or, if your statistical package supports it, a Cramér coefficient, page 153.*

50 There are two variables   51

There are more than two variables   52

51 Both sets of data are continuous (have more than 30 values) and are approximately normally distributed (a good way to get a feel

for this is to produce a scatterplot which should produce a circle or ellipse of points): *carry out a Pearson's product–moment correlation, page 154 (coefficient is* r*). This is the standard correlation method.*

Data are discrete or not normally distributed or you are unsure: *use a Spearman's rank-order correlation coefficient, page 158 or a Kendall's rank-order correlation coefficient, page 160. The marginal advantage of the former is that it is slightly easier to compare with the Pearson product–moment correlation as it is on a similar scale.*

Data are ranked: *use a Kendall's rank-order correlation coefficient, page 160. (The Spearman's correlation is marginally inferior in this case.)*

52 (Note: partial and multiple correlations are difficult to interpret.)

All sets of data are continuous and approximately normally distributed, you are interested in the direct level of association between pairs of variables: *use pair-wise measures of association using a Pearson's correlation, page 154.*

All sets of data are continuous and approximately normally distributed, you are interested in the overall pattern of association: *use partial correlation, page 171, which looks at the correlation between two variables while the others are held constant. (Multiple correlation is a possibility but is rarely supported in packages. Its disadvantage is in interpretation and its inability to distinguish positive and negative relationships.)*

Above does not apply or you are cautious: *carry out Kendall's partial rank-order correlation coefficient, page 171, a test that finds the correlation between two variables while a third is held constant. This may not be supported by your package. If it is not, pair-wise testing is the only alternative.*

53 The dependent variable is discrete or not normally distributed or ranked. Be warned that nonparametric regression is required and that this is rarely available in a statistical package.   54

The dependent 'effect' variable is continuous and at least approximately normally distributed with the same variation in 'effect' for any given value of the 'cause' variable. (There will often be a requirement for a transformation of the data. Proportions and percentages can be transformed using probits. Other distributions may be normalized using reciprocal transformations, logits or many other possibilities. It is important that efforts are made to fulfil the requirements for approximately normal data with equal variance using transformations.)   55

54 There is one independent 'cause' variable and one dependent 'effect' variable: *use Kendall's robust line-fit method. If this is not*   (55)

available consider reframing (usually by simplifying) your hypothesis somewhat to fit a nonparametric correlation. The only other alternative is to continue to a parametric test (55), being very cautious with interpretation of the results.

All other designs: *there is no satisfactory nonparametric test and certainly nothing in a statistical package yet. Either reframe the hypothesis or go to 55 and continue with a parametric test. If there are two 'cause' variables and one 'effect' then the 'cause' variables might be divided into a small number of categories (e.g. 'low', 'medium' and 'high') and then a Scheirer–Ray–Hare test carried out, page 131.* (55)

55 There is one dependent variable ('effect') and one independent variable ('cause'). 56

There is one or more dependent variable ('effect') and two or more independent variables. 58

The data for the dependent variable can be classified into more than one group (e.g. by 'species' or 'sex'). There is a variable that may affect the dependent variable: *ANCOVA (analysis of covariance) is required, page 172. This is a technique where the confounding variable, known as the covariate, is factored out by the analysis allowing comparison of the groups. Complex designs are possible but the most common is analogous to a one-way ANOVA with the data (e.g. dry weight) in classes (e.g. cultivars) and a variable known to be confounding factored out as the covariate (e.g. degree days).*

56 The independent 'cause' variable is measured without error. 57

There is known to be some measurement error associated with the independent variable: *model II regression is required, page 170. This is a rarely used technique and only occasionally appears in statistical packages. It has the odd property of always over-estimating the slope of the relationship compared to the result from normal (model I) regression, page 163.*

57 (As the theoretical shape of the relationship is often unknown the usual strategy here is to try both methods and see which gives the better fit.)

The relationship is likely to be a straight line or you are not sure of the form of the relationship: *linear regression (a.k.a. model I regression). [Note: in many cases the independent variable can be transformed to straighten the relationship between cause and effect (e.g. if the independent variable is 'size' and is right skewed then a log transformation will often improve a linear fit).]*

The relationship is curvilinear or complex: *polynomial regression or quadratic regression (a special case of polynomial regression), page 170.*

58 There is one dependent 'effect' variable and two or more independent 'cause' variables.
There are several 'cause' and 'effect' variables: *path analysis,
page 176.*

59 Your primary aim is to find the 'cause' variable(s) that are the
predictors of the 'effect' variable: *use stepwise regression, page
175.*
You want to establish a model using all the 'cause' variables: *use
multiple regression, page 175.*
(The distinction between these two is rather arbitrary.)

60 You have arrived at principal component analysis, discriminant
function analysis and other multivariate techniques for exploring
your data. The usual result of this type of exploration is to identify
simple relationships hidden in the mass of the data.
Some of these tests are described in Chapter 9.
There are several observed variables that are approximately
continuous and you have no preconceived notion about division
into groups: *use principal component analysis, page 177.*
There are a variety of variables that may be a combination of
'causes' and 'effects': *use path analysis, page 176.*
There are two or more sets of observations and one or more
grouping variables: *use MANOVA (multivariate analysis of
variance), page 185.*
There are two or more sets of observations, one or more grouping
variables and a recorded variable that is known to affect the
observed variables (e.g. temperature): *use MANCOVA (multivariate
analysis of covariance), page 185.*
There are several observed variables for each individual that
are approximately continuous and individuals have already been
assigned to groups (e.g. species): *use canonical variate analysis,
page 182.*
There are several observed variables for each individual that are
approximately continuous, individuals have already been assigned
to groups (e.g. species) and the intention is to assign further
individuals to appropriate groups: *use discriminant function
analysis, page 185.*
There are several observed variables for each individual and you
wish to determine which individuals are most similar to which:
use cluster analysis, page 185.
You have data on the relative abundance of species from various
sites and wish to determine similarities between sites: *use
TWINSPAN, page 187.*

4: Hypothesis testing, sampling and experimental design

This chapter expands on some of the ideas introduced in Chapter 2.

Hypothesis testing

Much of scientific investigation is based on the idea of hypothesis testing. The idea is that you formalize a **hypothesis** (H_1) into a statement such as 'male and female shrimps are different sizes', collect appropriate data and then use the statistics to determine whether the hypothesis is true or not.

However, it is not quite as simple as that. The statistical tests do not give a simple answer of true or not. First you have to realize that every hypothesis will have an associated **null hypothesis** (H_0); most statistical tests use the null hypothesis.

So, for this example, the hypothesis (H_1) is: 'male and female shrimps are different sizes' and the associated null hypothesis (H_0) is: 'male and female shrimps are *not* different sizes'.

What a statistical test determines is the probability that the null hypothesis is true (called the *P*-value). If the probability is low then the null hypothesis is rejected and the original hypothesis accepted.

Acceptable errors

In reality the null hypothesis is either true or false. Unfortunately, as the statistical test only gives a *probability* of the null hypothesis being true, there are two ways of making the wrong inference from the test. These two types of errors are usually called **Type I** and **Type II** errors.

		Null hypothesis	
		Accepted	Rejected
Null hypothesis	True	Correct	**Type I Error**
	False	**Type II Error**	Correct

In a **Type I** error the null hypothesis is really true (male and female shrimps are not different sizes) but the statistical test has led you to believe that it is false (there is a difference in size). This type of error is very dangerous and could be seen as a 'false positive'.

In a **Type II** error the null hypothesis is really false (male and females are really different sizes) but the test has not picked up this difference. Small sample sizes will often lead to a type II error. This type of error is probably

[22]

less dangerous than the type I but should still be avoided if possible.

The ideal statistical test should have an equal, and hopefully very low, chance of the two types of errors. A test which increases the chance of getting a type II error while decreasing the chance of a type I is said to be 'conservative' while one that increases the chance of a type I error is said to be 'liberal'. Although it is best to achieve this balance of type I and type II errors a cautious approach is to err towards 'conservative' tests.

P-values

Errors are the inevitable consequence of results based on probability. The lower the probability (*P*-value) the more confident you can be in the rejection of the null hypothesis but you can never be totally sure, unless you have measured the whole population, that you are correct. It is a usual convention in biology to use a critical *P*-value of 0.05. This means that the probability of a null hypothesis being true is 0.05 (5% or 1 in 20). Whenever a statistical test gives a result with a *P*-value less than 0.05 we reject the null hypothesis and accept the alternative hypothesis.

There is nothing magical about $P < 0.05$ it is just a convention. If you use a lower critical *P*-value then the chance of making a type II error is increased. If you choose a higher critical *P*-value then you increase the chance of making a type I error.

It is worth pointing out that if a *P*-value is less than 0.05 it does not *prove* that the null hypothesis is false it just indicates that it is unlikely to be true. Indeed statistics can never *prove* anything, they can only suggest that a hypothesis is very likely to be true or untrue.

Sampling

In almost all biological studies it will be impossible to account for every individual in a population. Therefore it is necessary to examine a sub-group of the total population and extrapolate from this to the whole population. The process by which the sub-group of the population is selected is sampling.

Nearly all statistical tests make a fundamental assumption that sampling of individuals will be at random from all the individuals that could possibly be sampled. This sounds simple but achieving a random sample is not always easy!

If a population is evenly distributed through a habitat then a single small sample would be enough to gain a good estimate of whatever it is you are interested in (e.g. total population size, mean age or weight). However, this is rarely the case and most populations have distributions that are either random or clumped. In such populations a single sample is unlikely to produce a good estimate of population size, or mean height, or the variance of leaf thickness.

There are a wide variety of sampling strategies in use. It is important to

[23]

choose a strategy that is appropriate to the population being investigated. There are several steps in the development of a sampling strategy.

Choice of sample unit

A sampling unit may either be defined arbitrarily, such as a quadrat or pitfall trap, or be defined naturally, such as leaf or individual. Usually, naturally defined sample units will be obvious but the choice of an arbitrary unit size may be important. If the unit is a quadrat then there will obviously be a trade-off in effort between the number of sample units that can be observed and their size, simply because it takes more time to get information from a large quadrat.

A sample unit might also be a length of time (e.g. if you are investigating pollination then number of times a flower is visited in a series of time periods of set length might be your data set).

> The size of sample unit will usually remain constant but may sometimes be variable, especially if the characteristics of a population distribution are being investigated. Methods using variable sample unit size to investigate distribution pattern are called quadrat-variance methods. These methods allow an observer to gain an insight into patterns of distribution in space or time by analysing the different characteristics of the samples (e.g. mean and variance) using different sample units.

Number of sample units

This is nearly always determined by the amount of labour available; the more time and people that are available the more information can be collected. However, it is possible to calculate the number of sample units required to produce an accurate estimate of the population size. In general, more sample units will be preferred as the number of sample units increases the accuracy of statistical tests. However, quantity should not be increased at the expense of quality. Poor quality data will have more inherent error and, therefore, make the statistics less powerful.

If you require general advice on the number of observations to make then I can only suggest that, as a rule of thumb, you need at least 20 observations for a sample using a measured variable and many more than that if the variable is a simple categorical one such as blood group.

Positioning of sample units to achieve a random sample

An unbiased estimate of a population is only possible if the sample units are representative of the total population. The easiest way of achieving this is for each sample unit to contain a random sample of the population under investigation. If quadrats are being used, then their position within the area under investigation should be chosen using random numbers to generate two coordinates that are then used to position a corner of the quadrat. Although this method of choosing a position using random numbers often requires an area to be marked out, it is to be preferred over the quasi-random techniques, such as throwing a quadrat, that are certain to introduce some involuntary observer bias.

Random walks are another way to sample at random without requiring the area to be marked out. The observer walks a number of paces determined by a random number and then makes an observation or places a quadrat. Then another random walk is taken before the next observation, and so on. The advantages of this method are that sampling can be very rapid and that it requires little preparation. The drawback is that this type of sampling may be severely biased by the observer.

True random samples are ideal in a perfectly homogeneous habitat, but in a heterogeneous habitat they are likely to produce a biased sample with an estimated variance greater than that of the total population. A simple method used to minimize this problem is to take a *stratified random sample*. The method is simple: the total area is divided into equal plots and an even number of sample units is taken at random from each plot. It is possible to divide the total area into plots of different sizes if there are known to be different habitats in the total area. In this second case the number of sample units from each plot should be proportional to its size.

It might be tempting to conduct systematic sampling with sample units placed at regular intervals across a study area. There is a statistical problem with this strategy as most statistics require that a sample is taken at random from a population. However, some field ecologists suggest that estimates derived from systematic sampling are, on average, better than those from random sampling.

Timing of sampling

Most populations will be affected by season, time of day and local weather conditions. It is very important that timing is taken into account either by sampling strategy or by later analysis.

What I have been considering here is the problem faced by an ecologist working in the field and trying to design a suitable sampling strategy. The use of the very powerful statistical technique; analysis of variance (ANOVA) is more common in the situation of a controlled experiment where you are analysing the effects of different levels of a treatment (e.g. concentrations of fertilizer or temperature) on some measured aspect of a population. Then, to get a true estimate of the effect of the treatment, experimental design will be of paramount importance.

Experimental design

I do not intend to say very much about experimental design here as there are whole books dedicated to the subject. However, that should not imply that experimental design is an uninteresting or unimportant subject. The appropriate design of an experiment is the key to successful analysis of a problem, for without the correct design you will never have the right sort of data. The problems of sampling still exist in a designed experiment, but the control of the system allows the experimenter to ensure that there are sufficient individuals to sample, that all factor combinations have the same number of observations.

Control

The idea of control in experiments is to remove the effect of all other factors apart from the one which is being investigated. The word is often applied to a group that has not been altered by an experimental manipulation and is used to compare with a group that has. The assumption is that everything apart from the manipulation is the same in the two groups so any differences must be the result of the manipulation. There are different ways of applying controls and different types of control. It is important to consider whether the control used in an experiment is adequate to convince a sceptical reader that the effect 'proved' by the statistics is real or not. As a general rule of thumb more control is required! It is always tempting to focus on the more interesting manipulated groups and not give enough attention to control.

Procedural controls

Procedural controls are often overlooked in experimental designs. The idea is that everything done to the manipulated group apart from the actual treatment is done to a procedural control group which is the same size as the experimental group and the untouched control group. The idea of procedural control groups has been widely used in recent medical studies and shown some interesting results. For example, a new supplementary treatment for a common disease is being investigated. Everyone in the study is given the conventional treatment. Each individual is then randomly (and secretly) assigned to one of three categories: the control group is given nothing else, the experimental group is told about the supplementary treatment and given the new drug and the procedural control group is treated exactly as the experimental group except they are given a placebo dummy drug (i.e. a chalk tablet or water injection). In this way the effect of the drug can be differentiated from the effect of the procedure.

Procedural controls are especially important if the experimental system requires a lot of preparation through building exclosure fences or with repeated visits to a site or many interventions in a laboratory population. The

[26]

technique should nearly always be used in conjunction with the 'untouched' control.

Temporal control

Temporal control is another aspect of experimental design that is worth incorporating. If the effect of a long-term manipulation is to be considered and there is only one control group and one experimental group available it is better to start the manipulation *after* the monitoring process has been underway for some time. The reason for this is that the differences between the two populations *without* any manipulation should be accounted for before the differences following the manipulation are tested. The ideal experiment will use half of the time for before and half for after manipulation. For example, if there are two lakes available to study the effect of eutrophication (surplus of nutrients) then the best design for a 2-year study is to monitor the lakes untouched for the first year and then to add nutrients to one of the two lakes during the second year.

Experimental control

Experimental control is any control of environment imposed by the experimenter. This is the classic type of control and is properly employed to remove all possible effects on the observations other than that from the experimental manipulation itself. Control as many factors as possible is always the best advice. So, if the effect of CO_2 levels on plant growth is being investigated then the experiment should control all the factors that may affect growth: light, temperature, humidity, water availability, soil organisms, soil type and nutrient availability. The degree of environmental control required to isolate the effect of the one factor being manipulated often leads to a very artificial situation with organisms being kept in isolation in perfect conditions. These controlled environments are often so far removed from the real world that the results are not really very informative.

Experimental control can be very expensive requiring growth cabinets, controlled temperature chambers or incubators for even rather simple investigations.

Statistical control

Statistical control is an alternative to experimental control. Instead of fixing all the possible factors that can affect the observations they are measured instead. Careful recording of all the environmental conditions, both biotic and abiotic, that are known to affect the observations being collected can then be used in statistical analysis of the data. Providing the experiment is not confounded (e.g. if all the manipulated individuals are in a cold area and all the unmanipulated in the warm) it is often possible to unpick the various effects and remove them from the analysis to leave only the effect of the manipulation. If statistical control is to be attempted then efforts should be made to ensure that adequate monitoring of all the possible effects is carried out and that the individuals in experimental and control groups experience a range of conditions.

Statistical control is usually cheaper than experimental control but requires more effort on the part of the researcher.

Some standard experimental designs

The **Latin square** is a system for placing replicates of treatments so that each of the treatment levels experiences each column and row of the experimental area. The reason for doing this is to avoid confounding the effect of the experimental treatment with any other factor that might be present in the experimental area (e.g. a gradient of soil quality). The arrangements for the four treatments suggested here is just one of many possible arrangements. Any arrangement of treatments such that each appears once in each column and row is OK although it is probably best, as here, to have each treatment level only occurring in one corner.

A	B	C	D
C	A	D	B
D	C	B	A
B	D	A	C

If the experiment is carried out in a series of locations (often called blocks in statistical jargon) it is important to ensure that each of the treatment levels is equally represented in each of the blocks otherwise any difference in conditions will be confounded with the treatment levels being investigated. Furthermore the position of the treatment levels within the blocks must not be repeated.

Block 1

A	C
D	B

Block 2

B	A
C	D

Block 3

D	B
C	A

Block 4

C	D
A	B

[28]

If a large number of samples are to be assigned to different treatment levels there are three obvious ways of assigning the levels. First to do all of level one then two and so on, second to carry out the assignments entirely at random and third to keep cycling through the levels in sequence. Each of these has problems. The first method will confound any external changes with the different treatment levels (e.g. if the experimenter becomes more efficient during the process). The second method is appealing but often leads to unwanted runs of the same treatment or too few replicates of particular treatments. The third may also confound the treatment levels with an external influence. The best strategy is a combination of the second and third methods and is called **stratified random** assignment. The assignments are made in batches with each treatment level appearing an equal number of times in the batch (usually one or two) but assigned at random. In the example shown below there are three treatment levels (X, Y and Z) assigned twice each in five batches of six.

Batch 1	Batch 2	Batch 3	Batch 4	Batch 5
X	Z	Y	X	Z
Z	X	Y	Z	X
X	Y	Z	Y	X
Y	X	X	Z	Z
Y	Z	Z	Y	Y
Z	Y	X	X	Y

5: Statistics, variables and distributions

There are many books available that discuss the history, philosophy and workings of statistics at length. That is not my purpose here, but it is important to have at least some idea of what statistics are, how different statistics are appropriate in different circumstances and that there are different types of data that you might collect. This chapter covers much of the same ground as Chapter 2 but in much more detail. However, I'm still only scratching the surface here and this section should only be used as a set of notes or pointers to further investigation of these subjects.

What are statistics?

In biology we are often concerned with groups of individuals. These 'individuals' might be single insects but they could also be, for example, herds, or species or blood cells. In most cases it is totally impractical to measure every individual in the group or groups we are interested in. What we are forced to do instead is to take measurements from a subset of the group. We call these subsets of the whole group **samples**.

We can ask and answer questions about the groups by formulating hypotheses. A simple question could be: 'is species A bigger than species B?'. If we had access to data from all individuals in a group we could answer this type of question very easily. However, we only have the sample and from the sample we have to extrapolate to the whole group. This is the job of statistics.

For example, if the hypothesis is that the mean sizes of populations of pike in two lakes are different we could easily find the answer if we had measured all the fish. However, in reality we only have a sample of 20 fish from each lake and the means of those samples might not be the same as that of all the fish. We carry out statistics on the information we have in the two samples to determine the probability that the hypothesis, or more usually its associated null hypothesis, is true. The idea of hypothesis formulation and testing has already been discussed in Chapter 4.

Types of statistics

I intend to say as little as possible about types of statistics here. However I feel it is important to give a feel for the differences and the way that statistics have been traditionally divided.

Descriptive statistics are usually the first to be calculated. They give information about the data you have collected. This can be a measure of the 'position' of the data, i.e. mean or median, and the 'dispersion', i.e. how variable the set of data is. Some descriptive statistics such as means (averages) will be familiar to everyone.

There is a division of statistics into two groups that are usually labelled 'parametric' and 'nonparametric'. This distinction is very real for statisticians but for those of us just using the tests it seems rather artificial.

Parametric statistics make assumptions about the form of the data under investigation. For instance, they usually require variables to follow known distributions, usually the normal. If the data do conform to the assumptions then these tests are usually more powerful and should therefore be preferred. There are also types of questions that can only be answered if assumptions about distributions are made.

Nonparametric statistics require little or no knowledge of the distribution. Therefore they are often called 'distribution-free', 'ranked' or 'ranking' tests. In general these tests are less powerful but 'safer' if you have not tested all the assumptions for a parametric test. Nonparametric tests are also somewhat restrictive and cannot be used to answer some more complicated questions.

In this book, unlike many other books, the chapters are not ordered according to the type of statistics. I have used the type of question you want to ask as the method of dividing up the book.

What is a variable?

To carry out any statistics you need some data to work with. First you decide what it is you are interested in and then select a suitable variable. The variable is the property that you measure. It is the 'food' of the statistics and choosing variables is something you must get right. For example, if you are interested in the occurrence of a scale insect on two strains of citrus trees then a suitable variable might be; 'number of scale insects on a leaf'. However, if the strains differ in leaf density or size of leaf then a more appropriate variable could be 'number of scale insects per cm^2 of leaf'. Then, the insects may be bigger on one strain than the other so perhaps 'mass of scale insects per cm^2 of leaf'. Each time the variable is refined in this example it becomes more difficult to obtain, taking more time and effort. There is a trade-off. More effort required for each observation leads to less data in total so any refinements to the variable collected must be warranted. Choosing the best variable is something of an art.

It is important to ensure that the variable or variables you choose to measure or collect are appropriate to the task.

Note: I use the term variable throughout this book as it is the one in common usage, although the correct term is variate.

Types of variables or scales of measurement

There are many types of variable.

Measurement variables

Measurement variables are variables where a numerical value is assigned. They can be further subdivided.

Continuous variables (sometimes called interval variables) — this type of variable theoretically has an infinite number of values between any two points. Of course, in practice the accuracy of measurement will not be perfect, as it will be limited by the observer and the equipment used. Therefore, there will only be a limited number of possible values between any two points. Obvious examples of continuous variables are lengths, weights and areas.

Note: two words that are often confused:

Accuracy is the closeness to the real value. This is usually set by the observer or the equipment and should be chosen as appropriate to the variable. When you write down a value it should reflect the accuracy with which the measurement was taken. If you measure to the nearest 0.1 g then 5 g should be written as 5.0 g, not 5.00 g;

Precision is the closeness of repeated measures to the same value. It is possible to have data that are very precise but very inaccurate. For example, your balance gives exactly the same value for repeated measures of the same object but they are all overweight because the balance was not calibrated properly. The data obtained would be precise but inaccurate.

Discrete variables (also called 'discontinuous' or occasionally 'meristic') — unlike continuous variables this type of variable has a limited number of possible values. These possibilities are often, but not always, integers. For example, number of live-born offspring in a litter of mice can only ever be an integer as there is no possibility of recording a fraction of an offspring.

Discrete variables are often produced by questionnaires. Respondents are offered choices such as: 1, strongly disagree; 2, slightly disagree; 3, neutral; 4, slightly agree; 5, strongly agree. There is clearly a continuous variable ('agreement') here and division of responses into categories in this way is rather arbitrary. It would be very easy to devise different ways of dividing the responses to obtain more or fewer possibilities.

The distinction between discrete and continuous variables can be rather blurred.

Example 1: a discrete variable becomes continuous. If you measure the number of cells in 1 ml of blood this must be an integer and therefore discrete. However, it has so many possible values that it is effectively continuous.

Example 2: a continuous variable becomes discrete. Seed diameter is a continuous variable but if you measure poppy seed diameter to the nearest 0.05 mm there will be only a few possible values making it effectively discrete.

How accurate do I need to be?

It is often possible to use better equipment or become more careful when measuring to increase accuracy. However, increased accuracy will take longer and result in less data being collected: another trade off. As a rule of thumb, there should usually be between 30 and 300 possible intervals between the smallest and the largest value. If possible, adjust the accuracy of the measurement accordingly. Don't assume that measuring to as many decimal places as possible will make the data any better.

Ranked variables

Ranked variables are data ordered by magnitude, although exact values are not relevant. It is not assumed that the difference between 1 and 2 is the same as that between 3 and 4. Often, it is possible to put observations into rank order without measuring at all. For example, plants from six pots could be ranked in 'health' order by simple observation and assigned values from 1 to 6.

Attributes (also called 'categorical' or 'nominal' variables) — usually 'yes' or 'no', 'male' and 'female' or a small number of possibilities. For example you could score flower colours as red, blue or yellow. Attributes or categories should, ideally, not have any obvious sequence.

Derived (or 'computed') — usually calculated from two (or more) other variables, e.g. ratios, percentages, indices, rates.

Warning: You lose accuracy by combining variables into ratios
For example, if we round to 0.1 then 1.2 implies 1.15–1.25 giving a maximum error of 4.2% and 1.8 implies 1.75–1.85 with a maximum error of 2.8%.

If these two observations are combined into a single observation then 1.2/1.8 implies a range from 1.15/1.85 to 1.25/1.75 or 0.622 to 0.714, giving a maximum error of 7%, which is much greater than the error in the original data.

Furthermore, 1.2/1.8 and 0.6/0.9 are indistinguishable.

Distributions of combined variables are often awkward. Be very careful with percentage data as percentages will often have rectangular (flat) distributions and/or have limits at 0% and 100%. However, percentages are a familiar and widely used method for expressing observations and there are a few statistical tricks available to help you deal with them. Ratios can also lead to a loss of information. For example both 5/10 and 500/1000 will give a ratio of 0.5, losing all information about the size of the sample.

Types of distributions

Why do you need to know about distributions?

Just as there are different types of variable, there are different types of distribution. All parametric statistics and many nonparametric ones are based on features of distributions or on assumptions about data following certain distributions.

Discrete distributions

The Poisson distribution

This is a very useful tool to use as a starting point in many biological investigations. It is a distribution describing the number of times an event occurs in a unit of time or space. Usually a sample of time or space is taken and the number of events recorded. Typical events are, for example, number of fish-lice on a fish or number of influenza cases reported in a week.

Assumptions of the Poisson distribution
1 Mean number of occurrences is small relative to the maximum possible.
2 Occurrences of one event must be independent of others.
3 Occurrences are random.

The purpose of fitting to this distribution is to test for randomness or independence in either space or time. If number of scale insects on leaves fits a Poisson distribution then it can be assumed that the assumptions hold. Therefore, the occurrence of individual insects is unaffected by the presence of others, so we can infer that the scale insects arrive at random and that no leaf is 'full' of scale. If the distribution is significantly different from a Poisson then not all the assumptions hold and further investigation should follow.

Poisson distributions only require knowledge of the mean, as mean and variance are equal. This property is also very useful, as simple inspection of the mean and variance of observations in a sample will give you some idea of the form of a distribution.

If variance > mean then the population is more clumped (aggregated) than random. If variance < mean then it is more ordered (uniform) than random (see Fig. 5.1). Distributions may be described by simply quoting their variance/mean ratio.

The binomial distribution

The binomial distribution is a discrete distribution of number of events when there are two possible outcomes for each event the probability of each is constant. For example if the probability of each birth producing a female is

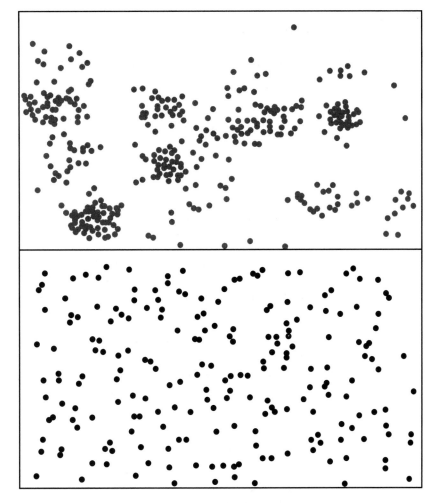

Fig. 5.1 Two hypothetical distributions of individuals in space. In the first the individuals are highly clumped or aggregated. If quadrats were used to sample from this population the variance in number of individuals per quadrat would exceed the mean. However, in the second distribution the individuals are more ordered than random and the results of number of individuals per quadrat would show a variance less than the mean.

0.5 (usually termed p) then the probability of a male is $1-0.5$ (also 0.5 in this case and often termed q), as there are no other possibilities.

This means that each individual being born has a 50% chance of being female and 50% chance of being male.

If this is expanded to families with more than one offspring then we can start to apply probabilities to the proportions of males (M) and females (F). For example, in a family with two offspring there are four possible outcomes: FF, FM, MF and MM (note that there are two routes to get one male and one female). As the chance of each event has already been determined as 0.5 then the chance of each of the four outcomes is 0.5×0.5 or 0.25. In other words there is a 25% chance of getting FF, 25% for MM and then 25% for each of MF and FM. So 50% of families with two offspring will have one of each sex.

[35]

This can be expanded further to three offspring where there are more possible families:

Female offspring	Male offspring	Probability
3	0	0.125
2	1	0.375
1	2	0.375
0	3	0.125

There are many uses of this expansion from single events to groups in biological investigation. To stay with the male–female example for the moment, an investigation into 480 broods of Song Thrushes where there were five eggs surviving to fledging gave frequencies (numbers of observations) for each of the six possible categories of families:

Females	Males	Probability	Expected no.	Observed no.
5	0	0.03125	15	21
4	1	0.15625	75	76
3	2	0.31250	150	138
2	3	0.31250	150	142
1	4	0.15625	75	80
0	5	0.03125	15	23

The expected frequencies from the assumption of a binomial distribution can be tested against the observed numbers using a chi-square test or a *G*-test. In this case, despite having fewer broods with three of one sex and two of the other than was expected, the difference is not significant, and therefore we accept the null hypothesis that the sexes of individuals in Song Thrush broods of five follow a binomial distribution with a p of 0.5 (i.e. there is a 50% chance of having a female offspring).

The binomial makes a very good starting place for a null hypothesis of even chances of events happening in all groups observed. If the binomial distribution is not followed then alternative explanations about aggregated or dispersed events have to be invoked.

Negative binomial distribution

In many organisms aggregation is almost ubiquitous. The negative binomial

[36]

distribution is a discrete distribution that can be used to describe clumped data (i.e. when there are more very crowded and more sparse observations than a Poisson distribution with the same mean). There are reasonable assumptions that can be made about the way organisms distribute themselves that result in a negative binomial distribution which allow sensible null hypotheses about aggregated distributions to be made.

Hypergeometric distribution

Another theoretical discrete distribution that has some use in biology is the hypergeometric distribution, which is used to describe events where individuals are removed from a population and not replaced. It is, therefore, quite useful in small, closed populations that are being sampled destructively and also in the application of mark–recapture techniques.

Continuous distributions

The rectangular distribution

(Also called a 'flat' or 'uniform' distribution.) This describes any distribution where all values are equally likely to occur. This distribution rarely appears in reality but it can sometimes be useful for generating a null hypothesis (see Chapter 4).

The normal distribution

The normal distribution is the most important distribution in statistics and it is often assumed that data are distributed in this way. Therefore, is it often important to determine whether the data set is a good fit to a normal distribution or not. Methods you can use to test this are the Kolmogorov–Smirnov test, the Anderson–Darling test or a χ^2 goodness of fit. This symmetrical, continuous distribution is described by two parameters: the mean, μ (describing the position) and the standard deviation, σ (describing the spread). It is sometimes called the Gaussian distribution. A normal distribution is symmetrical and always has a characteristic bell-shape.

In a perfect normal distribution

(μ is the mean and σ is the standard deviation)
$\mu \pm \sigma$ contains 68.25% of the observations: 50% fall between $\mu \pm 0.674\sigma$
$\mu \pm 2\sigma$ contains 95.45% of the observations: 95% fall between $\mu \pm 1.96\sigma$ (see Fig. 5.2)
$\mu \pm 3\sigma$ contains 99.73% of the observations: 99% fall between $\mu \pm 2.576\sigma$

The standardized normal distribution

This is a derived distribution where each observation in a normal distri-

[37]

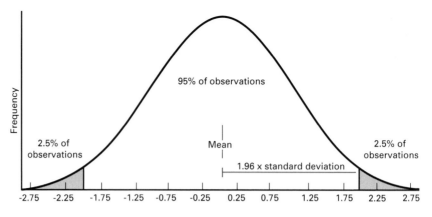

Fig. 5.2 In a normal distribution 95% of the observations will fall within 1.96 standard deviations of the mean. This leaves 2.5% of the observations in each of the tails of this symmetrical distribution.

bution is processed by subtracting the mean and dividing by the standard deviation. This gives a normal distribution with a mean of 0 and a variance of 1. The purpose of this transformation is to compare distributions that might have very different means on the same scale to look at the shape of the distribution.

Convergence of a Poisson distribution to a normal distribution

Even though a Poisson distribution is discrete (you can only get integers), when the mean number of observations is very large a Poisson distribution will approximate to a normal distribution. This could arise, for example, if you counted the number of springtails in a group of soil samples.

Note. A binomial distribution with > 100 observations (or fewer if $P \approx 0.5$) will also approximate to a normal distribution.

Sampling distributions and the 'central limit theorem'

The means of samples taken from *any shape* of parent distribution will themselves have a normal distribution. This is the basis for the rule that the standard deviation of the sample mean (i.e. standard error) of a sample is $\sigma/\sqrt{n}$, where σ = standard deviation of the observations and n = number of observations.

Describing the normal distribution further

Two types of departure from normality in a data set are:

Skewness. This is another word for asymmetry; skewness means that one tail of the bell-shaped curve is drawn out more than the other (see Figs 5.3 & 5.4). Skews are either to the right or left depending on whether the right or left tails

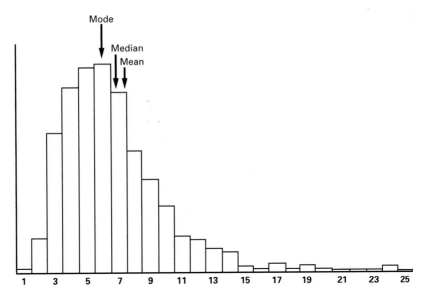

Fig. 5.3 This distribution is clearly right skewed and has a g_1 value well above zero. In a skewed distribution the mean is always nearer the tail than the mode with the median falling between mean and median.

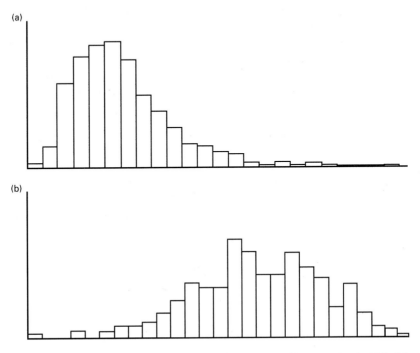

Fig. 5.4 These two frequency distributions are clearly not symmetrical. The data in (a) is right skewed and has a g_1 value of 1.53. The data in (b) is left skewed and has a g_1 value of –0.335.

[39]

are drawn out (i.e. long right tail, right skewed distribution). Statisticians label the true skewness parameter γ_1 (gamma$_1$) and the estimated value g_1. A negative g_1 indicates skewness to the left and a positive g_1 skewness to the right.

If a distribution is skewed the mean is nearer to the tail than the mode, as shown in Fig. 5.3.

Kurtosis. This is a measure of the 'flatness' of a distribution. A symmetrical distribution can differ from the normal in being either **leptokurtic** or **platy-kurtic**. A leptokurtic distribution has more observations very close to the mean and in the tails. A platykurtic distribution has more observations in the 'shoulders' and fewer around the mean and tails. A bimodal distribution is, therefore, extremely platykurtic. The kurtosis parameter is γ_2 (gamma$_2$) and is estimated by g_2.

In a perfect normal distribution both g_1 and g_2 are equal to zero. A negative g_2 indicates a platykurtic distribution and a positive g_2 leptokurtosis.

Is a distribution normal?

It is extremely unlikely that you will collect a data set that is perfectly normally distributed. What you need to know is whether the data set differs *significantly* from a normal distribution. One good way for checking data for departures from 'normality' is to use the **Kolmogorov–Smirnov (K–S)** test or the **Anderson–Darling (A–D) test**. Both tests compare two continuous distributions with the null hypothesis that the distributions are the same (i.e. it tests the sample data against a normal distribution with the same mean and variance). It is more powerful than the χ^2 goodness of fit method, which is another commonly used method of determining whether data set is normally distributed. See Chapter 7 for details of the K–S, A–D and χ^2 goodness of fit tests (pages 61–72).

Transformations

Parametric statistics assume that the data set you are using is distributed normally. So check that this is true first using a statistical test fitting your distribution to a perfect normal distribution with the same mean and variance. If the data set is significantly different from normal — try a transformation, such as: log; square-root; arcsin square root for percentage or proportion data; probits. There are many standard methods to try but as long as you treat each piece of data in exactly the same way you can do any transformation you like.

The *t*-distribution

This symmetrical, continuous distribution is related to the normal distribution but is flatter with extended tails. It is the distribution of deviations from the mean divided by the sample standard error of a huge number of samples. As the sample standard error varies between samples the spread is greater than if the deviations were divided by the true standard deviation of the mean (standard error). *t*-distributions have degrees of freedom (d.f.) associated with them that correspond to the size of the sample. So the smallest d.f. of 1 from just two observations will give the flattest distribution and a d.f. of ∞ will recapture the normal distribution.

Confidence intervals (CI)

Ninety-five per cent confidence intervals are calculated for samples using *t*-distributions. (Although when the true σ is known or the sample size is huge the normal distribution can be used.) Ninety-five per cent CI should be preferred to the more usually quoted mean ± SE, as the standard error (SE) of the mean is only really useful if the sample size if known and then it can be converted to a CI of the required width.

Be very careful when you see headings such as 'means and standard deviations', as this wording is slightly ambiguous. It usually translates as mean and σ of the observations but is, on occasion, referring to mean and its standard deviation (i.e. SE).

The chi-square distribution

This is another continuous distribution that is very useful in statistics. Unlike the normal and the *t* distributions it is asymmetric and varies from 0 to positive ∞. The chi-square distribution is related to variance.

X^2 is the usual way of expressing sample statistics approximating to χ^2.

The exponential distribution

The exponential distribution is a continuous distribution that is occasionally useful as a null model in biology. It occurs when there is a constant probability of birth, death, increase or decrease. So, for example, if a population of beetles invades a new area they may have an exponential increase in numbers as their rate of increase is constant. As soon as the population stops following the exponential distribution the rate of increase has clearly changed. This may indicate that intraspecific competition has reduced the growth rate or a predator is starting to have an effect. Exponential distributions can also be used to examine decreasing observations. For example, the amount of drug in the bloodstream after an injection may have an exponential decay with 10% being removed every hour. You can test an observed distribution against an expected exponential distribution using a variety of tests of difference.

Nonparametric 'distributions'

It is sometimes better to ignore distributions totally. This is the case when the data set is known to be awkward or difficult to transform. The advantages of making no assumptions about the distribution of the data are great as it allows greater flexibility but there are some limitations in the type of statistical tests that can be used and in the power of the tests.

Ranking, quartiles and the inter-quartile range

In nonparametric tests data sets are usually ranked before they can be examined statistically (computer packages do this for you). If a set of data is put in rank order from the smallest value to the largest then information about the position of the data set or the spread can be gained by inspecting values at certain points in the ranked data set (for example the median is the value of the data point in the middle of the ranked set).

A quartile is simply the value of the data point that lies a quarter of the way into a data set and it is commonly used to describe the spread of a nonparametric distribution.

The inter-quartile range is the difference in the values between the data point one quarter of the way down the ranked list to the point three quarters of the way down.

Box and whisker plots (box plots) summarize data where there are no assumptions of distribution. A sample is represented by a box whose top and bottom represent the upper and lower quartiles (i.e. the box covers the inter-quartile range). The box is divided at the median value. A line (the whisker) is drawn from the top of the box to the largest value within 1.5 inter-quartile ranges of the top and the same from the bottom. Any values outside this range are then added as symbols. These outliers are often identified in some way (they certainly are in the statistical package, SPSS) so you can check them as outliers are the values most likely to have been mistyped!

6: Descriptive and presentational techniques

These techniques are presented in roughly the same order as they appear in the key.

General advice. Descriptive and presentational techniques serve two rather different purposes. The first is to summarize and display data in the best way possible for a reader to derive information about the data. If this is the intention then the technique used should be as simple as possible and require the minimum effort from the reader. The second purpose is for researchers to explore their own data. A variety of methods should be employed that show the data from different perspectives. In this way you can become familiar with your data and may be stimulated to pursue new lines of enquiries or test different hypotheses.

This chapter is intended to offer general advice on data presentation and although all the examples are generated in the statistical packages featured in the next chapters there are no detailed descriptions for navigating the menus to generate the figures you see.

Displaying data: summarizing a single variable

Box and whisker plot (box plot) is an excellent way of summarizing data, especially if it is not normally distributed. The plot shows the median value as a thick bar, the interquartile range as a box and the full range as the 'whiskers'. Some statistics and graph drawing packages show outliers (data points well outside the range of others) as individuals points. An example is shown in Fig. 6.1.

Displaying data: showing the distribution of a single variable

It is important that any graphical depiction of data is clear. Usually the easiest and clearest way to display a single set of data is to use a histogram or a bar chart of frequency of occurrence. If you have discrete data then it may be best to display each possible value. However, in most cases it will be necessary to group the data into classes. There is often confusion about the difference between a histogram and a bar chart.

Bar Chart for discrete data. Each possibility is represented on the abscissa (x-axis), frequency on the ordinate (y-axis). Gaps between the bars symbolize the discrete nature of the data (see Fig. 6.2 for an example). If there are a very large number of possibilities then a bar chart may be inappropriate as clumping the data into groups will give a better picture of the distribution. If this happens then you have moved to a histogram.

[43]

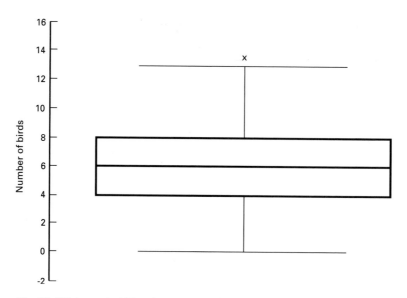

Fig. 6.1 This box and whisker plot was created in SPSS. There were 400 observations of numbers of bird species seen by a single observer from the same point during a fixed time. It shows that the median number was 6 and that 50% of the observations were between 4 and 8. Note that there was a single observation of 14 (marked as a cross) and that the axis extends to –2 even though 0 is clearly a lower limit (it is impossible to see fewer than zero species of birds).

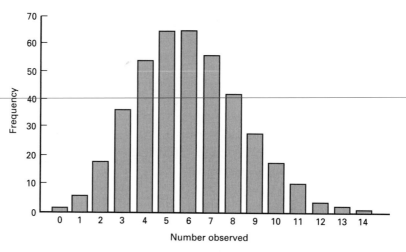

Fig. 6.2 This is the same data set as was used to create the box and whisker plot in Fig. 6.1. This SPSS chart shows the actual number of birds seen in a garden in a fifteen minute period. A total of 400 observations was made. Clearly the number of birds can never be lower than 0 although it might be greater than 14. Observations of this kind will always be integers although there is certainly no requirement for data to be integers to be suitable for bar charts. Gaps between the bars symbolize the discontinuous nature of the data.

Histogram for continuous data. Observations are grouped into artificial classes. The midpoint of the class is displayed as a label on the *x*-axis and frequency (number of observations) on the *y*-axis. No gaps should be left

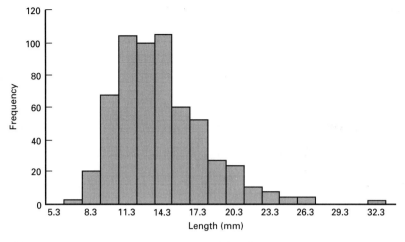

Fig. 6.3 This histogram presents a data set comprising 589 observations of elytra length in a population of beetles. The observations are linear measures and clearly continuous. All measurements were made to the nearest 0.1 mm. Each bar represents a range of values and there are no gaps between the bars. Values on the *x*-axis show the mid-points of the range for half the bars.

between classes to symbolize the continuous nature of the data. Shading, especially intense shading, should be used sparingly. See Fig. 6.3 for an example.

Number of classes to display in a histogram?

As a rule of thumb use 12 to 20 classes…

> But employ some common sense. Small samples should rarely need to have 12 classes and huge samples may be grouped into more than 20 classes.

As an alternative rule of thumb use $\sqrt{n}$ classes (where n is the number of observations in your sample).

> In the example of the beetle elytra in Fig. 6.3 there are 19 classes of 1.5 mm each, although only 15 have any observations. This fits with the first rule. The second rule of thumb suggests 24 classes.

Pie Chart for categorical data or attribute data. If the categories have no logical sequence (for example they are blood groups, species of tree or mutants of *Drosophila*) then a pie chart is probably a better method of presentation than a bar chart. However, if the categories have a logical sequence, such as five arbitrarily defined levels of ripeness, then a bar chart will be more informative. An example is shown in Fig. 6.4.

Tip do not use 3D bars or shadow effects on histograms or bar charts (unless it is for a display and then only in exceptional circumstances). Such effects obscure the data as it is difficult to see exactly where the top of the bar lies. I

[45]

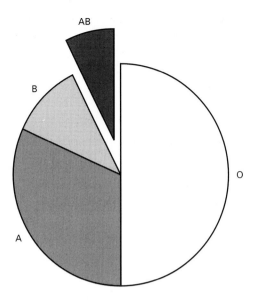

Fig. 6.4 This pie chart, generated in SPSS, shows the blood groups of a sample of 200 people. A pie chart is appropriate for this sort of data as if it was presented as a bar chart the *x*-axis would have no real meaning. Shading is not required but may be used if desired. Slices may be 'exploded' for emphasis as the slice AB has been in this example.

would also advise against the use of colour unless it is absolutely necessary (although I can't think of an example where I would advocate its use!).

Descriptive statistics

Statistics of location

There are several of ways of defining the 'location' of a distribution. It is tempting to focus only on the arithmetic mean as this is the easiest statistic to calculate and the most commonly used. However, it is worth considering some of the alternatives, especially the median.

Arithmetic mean. The 'normal' mean, often called an average and by far the most commonly used measure of location; when written it is usually denoted as **x-bar** (i.e. $\bar{x}$) which is an estimate of the true mean represented by the Greek letter, μ (mu) sometimes written as μ_x.

Geometric mean. The antilog of the mean of the logged data; it is always smaller than the arithmetic mean. The most commonly encountered use of this statistic is when data have been logged or when data sets that are known to be right skewed are being compared.

Harmonic mean. The reciprocal of the mean of the reciprocals and is always smaller than geometric mean. This type of mean is rarely needed.

[46]

Median. The middle value of a ranked data set. After the arithmetic mean it is the next most commonly used measure of location. It is the measure highlighted in box and whisker plots. If all the data are put into rank order (arranged in a list in from the largest value to the smallest) the median is the value associated with the middle ranked item (half-way down the list).

Mode. This is the most 'fashionable' value in a set of data; the value that occurs most frequently. It can be used with any type of data, even categorical.

> Variables with one clear mode are said to be unimodal. Distributions with two peaks are bimodal, more than two multimodal. The trough between two modes is sometimes called the antimode (Fig. 6.5).

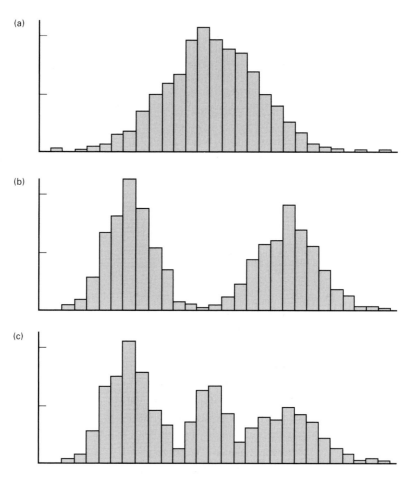

Fig. 6.5 Three rather different frequency distributions. In (a) there is a clear single mode of a unimodal distribution. In (b) there are two almost totally distinct distributions giving a bimodal distribution. This might indicate two separate populations, different genders or different species. In (c) the pattern of the frequency distribution is even more complex. There are three distinct modals making a multimodal distribution. This may indicate three cohorts of different years of recruitment in a population.

[47]

One of the problems with the use of the mode is that it is rarely suitable if the observations are made with any degree of precision (e.g. femur length to the nearest 0.01 mm) as there will be a much lower chance of an observation being repeated. Therefore, the mode should only be used when there are either a very large number of observations or a fairly small number of possible values.

Note: In any unimodal, symmetrical distribution (for example, a perfect normal distribution) the mean, median and mode are all the same (see Fig. 5.3 to see what happens in an asymmetric distribution).

Statistics of distribution, dispersion or spread

There are several ways to display the distribution or spread of a set of observations. However, it is important the measure used is appropriate to the data *and* the statistic of location (e.g. median) used.

Range. This is the most basic measure of dispersion and is simply the difference between the largest and smallest observations in a sample. It is usually quoted as the smallest and largest value (e.g. range $= 9.76-15.23$ cm).

Interquartile range. This is a nonparametric measure of dispersion that works on the ranked data. It is the difference between the value of the data item 25% of the way down a ranked list and the one 75% down. These values are called quartiles. The interquartile range is much more useful than the range as it is unaffected by outliers. Unlike many other measures of dispersion, the interquartile range is not necessarily symmetrical about the median. The quartiles are often given the codes 'Q1' and 'Q3'.

Variance. (Or mean square; estimate of sample variance s^2; true variance σ^2) mean of the squared deviations of observations from their arithmetic mean. Variance is rarely used as a descriptive statistic as it is not in the same units as the original observations. However, many statistical tests use variance in their calculations.

Standard deviation (SD). (Estimate of population standard deviation s; true SD σ) square-root of variance. This is commonly used as a descriptive statistic as it is in the same units as the original measurements or observations. However confidence intervals should be used if comparisons of different sets of observations are required.

Standard error (SE). By convention this is short for 'standard error of the mean' (i.e. the standard deviation of a distribution of means for repeated samples from a population). Standard errors are often quoted with means although this is probably because they are small rather than for any good statistical reason! If several samples are to be compared then the confidence interval should be preferred. If a measure of the variation in the sample is required then standard deviation is better.

> In theory there is no difference in calculation between a standard error and a standard deviation just that the former measures the standard deviation of a hypothetical sample of means.

Confidence intervals (CI) or confidence limits. These are derived from the standard error of the mean. Confidence intervals are the most useful measure of the dispersion of a distribution.

If a sample from a population is very large then the true mean of the population is 95% likely to lie within 1.96 standard errors of the sample mean. This region is called the 95% confidence interval, as you are 95% certain that it contains the true mean of a population. As samples get smaller then the multiplier used gets larger and the confidence intervals get wider. (If you have statistical tables it is easy to determine the required multiplier as it is derived from the *t*-distribution.) Confidence intervals are always symmetrical about the arithmetic mean. They are to be preferred over standard errors if several sets of observations are being compared.

Coefficient of variation. This is used to compare the amount of variation in populations with different means where direct comparisons of the standard deviations (s) are difficult to make as they are confounded by differences in scale. The coefficient of variation is usually denoted V or CV. $CV = (100s)/$ mean, and is usually expressed as a percentage.

Other summary statistics

There are other components of shape of the distribution of observations that can be interpreted easily. Knowledge of the skewness of a data set is particularly useful.

Skewness. This is a measure of the symmetry of a data set. If the data is symmetrical then the value of skewness will be 0. If there is a tail to the right it will be positive; if there is a tail to the left it will be negative. Meaningful values for skewness are only possible if there are more than 30 (and preferably a lot more) observations in the data set. Normal distributions are symmetrical and consequently have a skewness of 0. Skewness is discussed in Chapter 5.

Kurtosis. This is a measure of the shape of a distribution. It tells you whether there are more observations around the mean or less when compared to a normal distribution. Meaningful values are only possible if there are more than 100 observations in the data set.

Using the computer packages

General

All statistical packages will give summary statistics for sets of observations. However, generating exactly the set of statistics you are interested in may take several steps. The less frequently used statistics such as kurtosis may not be available.

SPSS

SPSS. In this package the data may appear in the same spreadsheet form as in a package such as Excel but the approach is rather different, as the statistics are displayed not on the spreadsheet but in a separate window. The data should be in a single column with an appropriate label. To change the column label simply double-click on the column name ('var0001' by default) and replace with something more suitable. The screen shot shows the various measures of dispersion that are available under the 'Descriptives...' options in the 'Summarize' submenu of the 'Statistics' menu. The default selections are shown which include the rarely useful minimum and maximum (Fig. 6.6). Once you have chosen the options you want the statistics are displayed in the '!Output' window. As you proceed through an SPSS session the nongraphic output accumulates in this window. This is very useful as you can go back and check results of previous tests very easily.

Further descriptive statistics can be accessed. From the 'Statistics' menu select 'Descriptives' then 'Frequencies...' selection. Select the 'Statistics...'

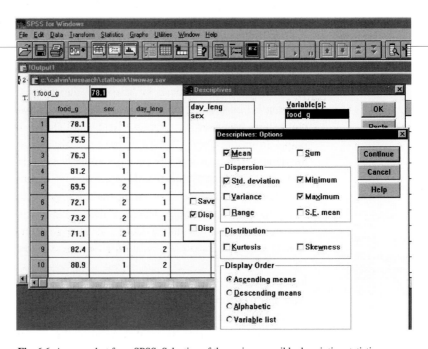

Fig. 6.6 A screenshot from SPSS. Selection of the various possible descriptive statistics.

button in the dialogue box and an array of options such as 'Skewness', 'Kurtosis', 'Mode' and 'Median' are available.

MINITAB

MINITAB. The data for a single variable should be in one column in the spreadsheet section of the package. The variable should be named appropriately in the cell under 'C1' although you are limited to eight characters. To get simple descriptive statistics go to the 'Stat' menu and select 'Basic statistics' and then 'Descriptive statistics...'. Move the name of the column with the data from the list on the left into the 'Variables:' box. Either leave the 'Display options' as tabular form or, if you want a more detailed output, change this to 'Graphical output'. Click 'OK'.

With 'tabular output' selected this is the sort of output that is generated:

```
Descriptive Statistics


Variable        N      Mean   Median   TrMean    StDev   SEMean
Height         24    11.338   11.335   11.337    0.088    0.018

Variable      Min      Max      Q1      Q3
Height     11.180   11.520   11.295   11.380
```

Data output 6.1

All the basics are reported here. The number of observations in the data set ('N'), the arithmetic mean ('Mean'), largest and smallest values ('Max' and 'Min'), standard deviation ('StDev') and standard error ('SEMean') as well as the 'nonparametric' versions of these statistics: median and the upper and lower quartiles ('Q3' and 'Q1'). Quartiles are explained further on page 48. If the 'Graphical output' is selected there is considerably more output to assess (shown in Fig. 6.7).

This output contains much of the same information as the 'Tabular' version but with some extras. In the mass of output on the right of the output the first thing is a test for normality. This is a test to determine whether the data in question is normally distributed. The 'A-squared' value is the output from a test and 'P-value' is the probability that the data is normally distributed. If the 'P-value' is less than 0.05 then this means it is unlikely to be normally distributed and therefore parametric statistics should not be used.

After this test comes the more usual descriptive statistics of arithmetic mean, standard deviation, variance and then the measures of the 'shape' of the distribution: skewness and kurtosis and the number of observations 'n of data'. Next comes some information about the data arranged in rank order. The value of the smallest and largest observations and then observations one-quarter (1st quartile), half (median) and three-quarters (3rd quartile) of the way down a ranked data list.

The last section gives 95% confidence intervals for three of the descriptive statistics. 'Mu' is the arithmetic mean, 'Sigma' the standard deviation and the median.

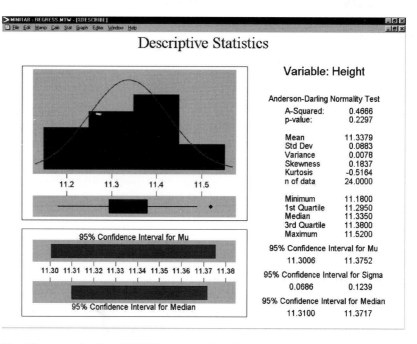

Fig. 6.7 A screenshot from MINITAB when the 'Graphical Output' version of descriptive statistics has been selected.

On the left of the output are four graphs. First is a histogram of the raw data with a normal distribution superimposed on it (the normal distribution shown has the same mean and standard deviation as the data). Then comes a box and whisker plot of the data (described earlier in this chapter) and finally graphical representations of the mean ('Mu') and median with their 95% confidence intervals.

Excel

Excel. In this package you have to assign a cell of the spreadsheet to contain the summary statistic you require. This is done by just clicking on an empty cell. Then you identify the cells that contain the variable (raw data) that you are interested in and the statistic appears. For example, your data, containing 100 observations, has been typed into the first column of the spreadsheet (column A). The first cell has the title of the variable and the actual observations are in rows 2 to 101. To calculate the arithmetic mean of this variable you go to any cell and declare its contents as '=AVERAGE(A2:A101)'. (As you can see Excel calls the arithmetic mean the 'average'.)

Other summary statistics are easily accessed using the 'function wizard' facility of the package. Or, once you have learned a few of the function codes you could just type them in. For instance '=STDEV(A2:A101)' to get the standard deviation reported in an empty cell.

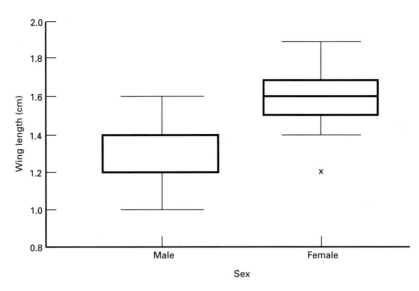

Fig. 6.8 In this SPSS-generated figure the sample of observations of wing lengths of a moth are divided into two groups by gender. As in most insects the females are considerably larger than the males and although there is some overlap in the 'whiskers' there is no overlap of the interquartile range of the two groups. Note that for the males the median and lower quartile are superimposed showing that 25% of the observations for males were almost of the same value.

Displaying data: summarizing two or more variables

Box and whisker (box plots). These are a good way of comparing two variables. They allow direct visual comparison of both the location and the dispersion of the data. An example of the use of two box plots is shown in Fig. 6.8.

Error bars and confidence intervals. A similar way of looking at the same data is to display the arithmetic mean and some measure of the dispersion of the data. An example of the use of mean and confidence interval is given in Fig. 6.9. Note that the interquartile range is not symmetrical about the median (Fig. 6.8), whereas the 95% confidence intervals (or standard deviation if you have chosen to display that instead) are symmetrical about the mean (Fig. 6.9).

You can display more than two groups using these methods. They provide a very powerful method of showing differences and similarities between many groups. In the example here there is almost no need for any further statistics —the difference between males and females is so striking!

Displaying data: comparing two variables

Associations

The best can be used if two observations are made for a single individual (e.g. the 'individual' is a stream and the water pH and stream flow have been

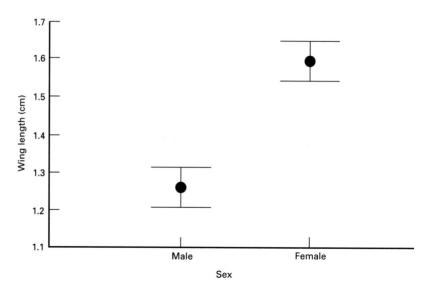

Fig. 6.9 This figure, also generated in SPSS, uses the same moth wing lengths as Fig. 6.8. The means for males and females are represented by filled circles and the 'whiskers' extend to cover the 95% confidence interval for the mean (i.e. there is a 95% chance that the true mean of the population lies between the extremes shown). There is no overlap of the 'whiskers' suggesting that the groups are likely to be highly significantly different.

recorded). Before any statistics are applied it is best to get a 'feel' for the observations by a graphical representation of the data.

Scatterplots. The simplest way to display a relationship between two variables is to use a plain scatterplot (Fig. 6.10). This assumes that two observations on the same row in the package are two measurements from the same 'individual'. An 'individual' can be almost anything: sampling station; greenhouse; pair or single bone.

It is important that all figures should have appropriate axis labels on them. They should also be accompanied by a figure legend that makes the plot interpretable without reading the relevant section of the text.

Do not add extra information that is not relevant or appropriate. For example many packages offer best-fit lines as a simple option. Do not use these unless: (1) you believe there is a 'cause' and 'effect' relationship between the variables; (2) you have used regression and you want a graphical accompaniment; (3) you intend to use regression; (4) you wish to use one variable to predict the other.

Multiple scatterplots. A good way to compare observations from two sites where the same variables have been recorded is to use a multiple scatterplot. The axes will be exactly the same as for the single scatterplot but each group will be displayed using a different symbol. This technique works particularly well for two or three groups and less well for more than that. Choose symbols carefully to make the groups easily distinguished and make sure that the figure caption makes it clear which symbol matches which group (Fig. 6.11).

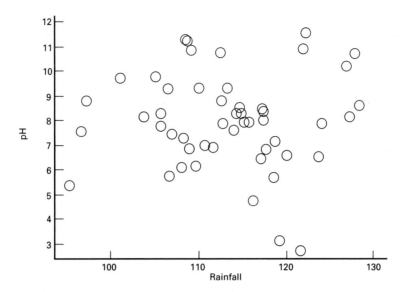

Fig. 6.10 The scatterplot of pH and rainfall from a range of sites shown here has been created in MINITAB using the default options. Clearly there is no obvious relationship between the two variables.

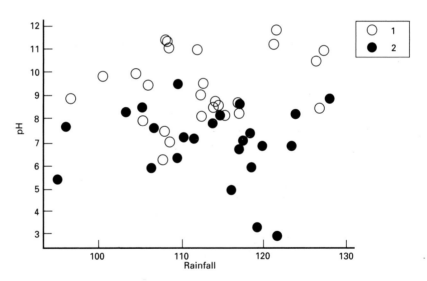

Fig. 6.11 In the example shown here two sets of observations from different study areas are identified with different symbols. A quick glance shows that group 1 is associated with a higher pH than group 2 but there is no obvious difference between the groups on the 'rainfall' axis. An analysis of variance or *t*-test could be used to determine the statistical probabilities but the results would only confirm what is obvious from the scatterplot.

More sophisticated use of symbols can convey a great deal of information about several factors on the same scatterplot. For example if the data for two morphological variables is collected and the individuals are divided into groups by sex and species then all this information can be incorporated in a

[55]

single plot. This can be done by using shaded and nonshaded symbols for the two sexes and two shapes for the two species.

Trends, predictions and time series

Lines. These should only be used to join points if there is a reasonable assumption that observations could be made between the points (Fig. 6.12). This is perfectly reasonable if the *x*-axis is temperature with readings made at 15–35 °C in steps of five as intermediate temperatures are valid. However, if the *x*-axis is number of eggs in a nest and the *y*-axis egg weight then it is perhaps unwise to draw a line linking mean weight at four eggs with mean weight at five as the line will pass though impossible points.

If there are several observations for each point on the *x*-axis then it is usually better to plot the mean or median with a measure of the dispersion at each point rather than use a scatter of points (Fig. 6.13). The same guidelines for joining means apply as for joining single observations.

It is very easy to deceive a reader by altering scales. For example if there is a slight but steady increase in the concentration of nitrate in a lake over time then this can be made to look like a rapid increase if the scale on the *y*-axis starts not from zero but from a value just below the lowest observed value. This kind of manipulation of the reader will work particularly well if there is no measure of the variation given either.

Fitted lines. The best way to draw the reader's eye to a relationship between

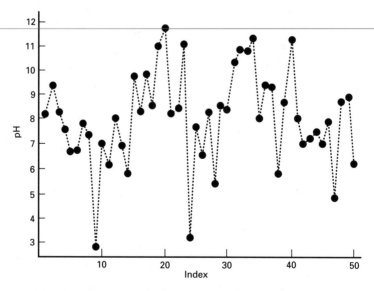

Fig. 6.12 This MINITAB-generated example of a line graphs shows a set of 50 readings of pH made through time in a chemical plant. The time gaps were equal and the observer thought it valid to join the reading made with lines as there is a reasonable expectation that the intervening times would have intermediate levels of pH.

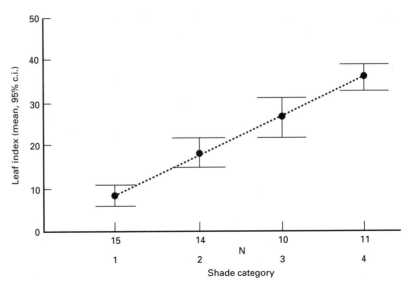

Fig. 6.13 This figure, generated in SPSS, shows a combination of a line graph and the mean and confidence interval approach of Fig. 6.9. In this case there are four levels of the variable shade that can be said to form a valid sequence from light to dark. Observations of leaf shape (a continuous variable) were taken at each of the four shade categories and the mean and 95% confidence intervals are plotted here with the means joined to emphasize the clear trend.

two variables is to use a fitted line of some sort. Indeed, an observer can sometimes be fooled into seeing a relationship in a scatterplot when there is none (Fig. 6.14). For this reason the use of fitted lines should be restricted to circumstances when the line is meaningful. The commonest use is to illustrate a relationship between two variables that has been investigated using regression. One technique is to plot the scattered observations along with the fitted line and then give more information about the regression in the text or the figure legend.

Confidence intervals. If a regression line has been calculated then it should always be displayed with its confidence intervals. This shows the range within which the line is 95% likely to lie (Fig. 6.15). If the confidence intervals are wide apart then the line is obviously less reliable.

Displaying data: comparing more than two variables

Associations

It may be tempting to use the full capacity of the graphics on the package you are using but there is little or nothing to be gained by plotting a multidimensional plot that is impressive to look at but impossible to interpret.

Three-dimensional scatterplots. This type of figure looks impressive but is quite difficult to interpret for several reasons associated with representing

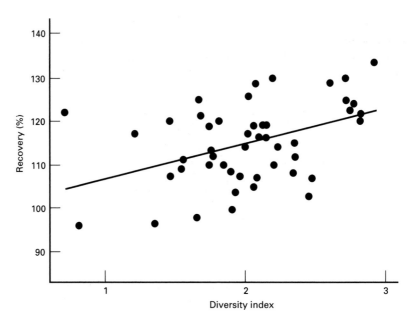

Fig. 6.14 This MINITAB-generated scatterplot of the recovery of biomass after an extreme event against the diversity index before the event shows a slight but nonsignificant trend. However, the addition of the trend line draws the eye and emphasizes the slight trend, convincing the reader that the trend is real.

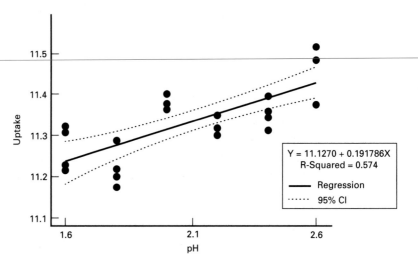

Fig. 6.15 Here the raw data, regression line and the 95% confidence intervals of the regression line are all shown along with some of the regression output from MINITAB. The variable 'uptake' measures the amount of drug passed across the stomach lining of a rabbit at various experimental pH levels. There is a clear relationship: the regression line slope is significantly different from zero and it explains 57.4% of the variation in the uptake observations. The 95% confidence intervals confirm that the relationship is robust. Note that the line, quite properly, does not extend beyond the data as predictions can only be safely made in the range of the data.

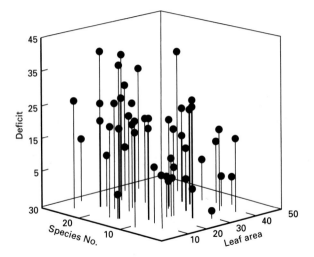

Fig. 6.16 Most statistical packages have them (this one is plotted using SPSS). The three-dimensional scatterplot is very difficult to interpret even when the relationship is quite strong and impossible when the relationship is weak. The spikes make the figure very cluttered but are vital to place the point accurately.

three dimensions in a two-dimensional medium. (1) If there are too many points plotted then those nearest the 'front' will obscure those as the back. (2) As the display medium is two-dimensional all the points need to have 'spikes' to anchor them to the $x=0$, $z=0$ plane. Without these spikes then a point near to the front but high on the y-axis will look *identical* to one near the back but low on the y-axis. (3) There is often no forced perspective making the arrangement of the axes seem odd and points at the back having the same size as those at the front fools the eye (Fig. 6.16).

Multiple trends, time series and predictions

Multiple fitted lines. Further information may be conveyed if two lines are fitted on the same graph. The advantage of this approach is that lines may be compared directly but the disadvantage is that the message may become confused. I would advise against a tactic I have seen used increasingly which is to have different y-axes for the same x-axis so that the two lines being compared fit sensibly. There are two problems with this type of graph. First it makes the reader see relationships that are not really there and second it is often difficult to see which scale applies to which line.

Surfaces. Many statistical packages include the option to have spectacular three-dimensional surface plots. I would advise against the use of these in almost all situations. The problems of all 'three-dimensional' graphs on two-dimensional surfaces apply with the additional problem that the solid or apparently solid surface totally obscures much of the surface. The way the points are connected to form the surface is questionable too: for example, if one or two of the axes have data that are normally distributed. This means that these observations are contributing a great deal of information in the centre of their range and rather less at the extremes. The surface plot does not reflect this in most cases (except it often betrays this by tending to have smoother edges where the surface is extrapolated from fewer points). Therefore, the edges of the surface can be influenced by the extreme points, the very points likely to be measured with less accuracy.

Remember that a surface, like a joined up line, can only be used if both the '*x*' and '*y*' (or '*z*') observations can reasonably be expected to have possible intermediate values (Fig. 6.17).

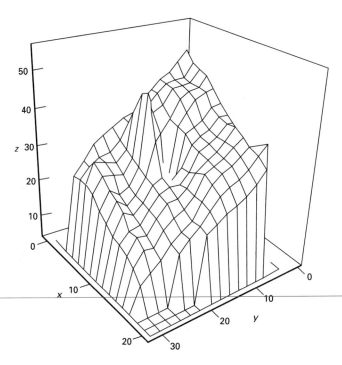

Fig. 6.17 Possibly even more difficult to interpret than the 3D scatterplot shown in Fig. 6.16 is the 3D surface plot such as this one drawn in MINITAB. This sort of figure only works if relationships are very strong or the smoothing algorithm is so strong that all the variation is wiped out of the data. Particular problems of 3D surfaces are that the edges tend to be extrapolated from far fewer data points than the middle and that peaks can obscure a lot of the surface behind them.

7: The tests 1: tests to look at differences

These are presented in roughly the order they appear in the key.

Do frequency distributions differ?

Questions

There are two basic types of questions that can be asked:
1 Does one observed set of frequencies differ from another?
2 Do the observed frequencies conform to a standard distribution?

In the first case the test becomes an analogue of a two-sample test of difference, such as the *t*-test. In the second it is a way of testing observations against expected frequencies, such as in plant breeding crosses when particular ratios of phenotypes are expected or to test if organisms are occurring at random by testing against the Poisson distribution. The *G*, chi-square goodness of fit, Kolmogorov–Smirnov (page 70) and Anderson–Darling (page 72) tests are the most commonly employed tests to answer these questions and are described below.

G-test

In situations where you have observed frequencies of various categories and expected proportions for those categories that were not derived from the data themselves then the *G*-test should be the preferred statistic to use. However, it is not many years since this test was shown to be superior to the traditional chi-square goodness of fit approach on theoretical grounds. Consequently, it is not supported by any of the packages considered in this book. If your package supports the *G*-test then use it and its associated correction factor, the Williams correction.

Chi-square test (χ^2)

Often known as the chi-square goodness of fit, this test is one of the most widely used in the whole of biology. It is also the statistical test you are most likely to be familiar with. You will usually present the data in a table showing the observed and expected frequencies for various categories. These categories can be single outcomes or groups of possible outcomes. It is customary to use grouping of categories to ensure that none of the expected values is less than 1 (some authors, erring on the side of caution, suggest 5). The expected values can either be derived from a distribution such as the Poisson or negative binomial, assume that all categories are equally likely (a flat or rectangular distribution) or they can be derived from another set of data. In all cases the

Chapter 7
The tests 1: tests
to look at
differences

null hypothesis (H_0) will be that the observed and expected frequencies are not different from each other. The chi-square test may also be used as a test of association (see next chapter).

An example

A very common starting point for investigations in biology is to determine whether events or observations are occurring at random. If events are truly random then they should follow a Poisson distribution (see Chapter 5 for more details). The chi-square test allows you to compare observed data with the expected data following a Poisson distribution with the same mean. In this case the number of lice found on adult char is recorded. All observations were taken from a single catch of 98 fish. If the lice attach themselves to the fish at random then they will follow a Poisson distribution and the chi-square will not be significant (i.e. null hypothesis is that lice attack at random). If the result is significant then lice do not attach randomly and a new hypothesis should be formulated.

No. lice per fish	0	1	2	3	4	5	6	7	8+
No. observations	37	32	16	9	2	0	1	1	0

SPSS

SPSS. First I should point out that processing this type of data for a goodness of fit chi-square in SPSS is not easy unless you wish to fit the distribution to a uniform one (i.e. all categories are expected to have the same number of observations). If you have this sort of data use the Kolmogorov–Smirnov (K–S) test to answer the question in SPSS. However, I will go through the procedure anyway assuming you don't know how to calculate Poisson 'expecteds' by hand (perhaps the faint hearted should move on to the K–S test now!).

1 Make sure that the data are in a single column with the actual data, not the frequencies. In this case there will be 98 rows in the data set, one for each fish. Label this column 'No_lice'.

2 Determine the mean number of lice/fish using the 'Statistics' menu, then 'Summarize' and selecting either 'Descriptives...' or 'Frequencies...' before moving 'No_lice' into the 'Variable(s)' box. The output will confirm the mean as 1.143 or 1.14.

The next steps show how to calculate the expected frequencies for a Poisson distribution with this mean in SPSS. If you can do this by hand or in a spreadsheet I suggest you do so and skip directly to step 8.

3 Generate a new column with all the expected frequencies in it (e.g. 0, 1, ... 7) in separate rows. Label this column 'freq'.

4 Use the 'Compute...' which is under the 'Transform' menu to bring up a dialogue box. In the 'Target variable' box enter, say 'exp1'. Then scroll down the list of functions until you reach CDF.POISSON(q,mean). (Note that CDF stands for cumulative density function.) Select this and move it into the

Chapter 7
The tests 1: tests
to look at
differences

'Numeric expression' box. Replace the word 'mean' with the mean you calculated in step 2 (i.e. 1.143 in the example). The 'q' should be replaced with the name of the variable you created in step 3 (i.e. 'freq'). Click 'OK'. A new variable will appear on the spreadsheet with numbers starting from 0.32. This first number is the probability of getting a zero in a Poisson distribution with a mean of 1.143.

5   Unfortunately the other values are cumulative values so the number in the second row is the probability of getting a 0 or a 1. The actual probability of getting a 1 can be calculated by hand using simple subtraction (i.e. the probability of getting a 1 is the probability of getting 0 or 1 minus the probability of getting a 0). The probability of getting exactly 2 is the difference between the cumulative value for 1 and 2. When your calculated probabilities drop below 0.01 (because with a sample of 98 this will drop you below the critical threshold of an expected value of 1 or more which is the limit for a chi-square test) move to step 8. If you have a high mean it may be easier to do another couple of steps in SPSS.

6   Repeat step 4 but use a different 'Target variable', say 'exp2' and replace 'q' with 'freq-1' instead of just 'freq' (assuming you labelled the list of numbers as 'freq'). This will produce a new column of probabilities with a zero at the top.

7   One final calculation step is required. Go to the 'Transform' menu and select 'Compute...' again. Insert a new label in the 'Target variable', say 'expected' as this is going to be the true expected value. Select 'exp1' from the list on the left and move it to the 'Numeric expression' box. Press '-' and then move 'exp2' over too. Put the expression 'exp1-exp2' in brackets then add '*98' (or whatever your sample size is if you are not using the example) outside the brackets. This will multiply the values by 98 turning your expected frequencies into expected numbers of lice.

8   You will notice that all expected frequencies for more than four lice/fish are less than one and should be grouped together to form a category of four and above. Write down the expected frequencies for the five categories.

No. lice/fish	0	1	2	3	4+
No. observations	37	32	16	9	4
Expected	31.3	35.7	20.4	7.8	2.8

9   As all values above 3 lice per fish have been grouped for expected frequencies, this must now be done for the actual data. In SPSS you can either do it by hand or use the 'Recode' option under the 'Transform' menu. Selecting either an 'Into same variables...' (destroying the original data) or 'Into different variables...'. If you choose the latter, move 'No_lice' into the 'Numeric variable -> output' box. Then type a name for your new variable

Chapter 7
The tests 1: tests
to look at
differences

in the 'Output variable, Name' box and click 'Change'. Your new name appears after the old name in the main box. Then select 'Old & New Values'. This brings up a bewildering set of options that are actually extremely useful in SPSS. In the 'Old value' section on the left select the button for 'Range' with a box and the word 'through highest' under it. Put a 4 in the box. Then in the 'New value' section on the right put a 4 in the 'value' box and click 'add'. This will put '4 thru highest -> 4' in the 'old -> new' section (meaning that all 4s or higher will be 4s). Finally click 'All other values' on the left, 'copy old values' on the right, and 'add' to keep the rest as they were. Click 'Continue' here and then 'OK' in the next window to create the new variable.

10 Finally choose 'statistics', 'nonparametric tests' and 'Chi squared' to bring up the 'Chi squared test window'. Move the variable you created in step 9 into the test variable box. Unfortunately you have to enter the expected frequencies one at a time. Click on values in the 'expected values' area and enter the expected frequencies, starting at the one for zero (i.e. 31.3 in the example) and clicking 'add' after each one before clicking 'OK' to run the test. This is the output you should expect.

```
- - - - - Chi-Square Test

    NEWLICE

                    Cases
    Category   Observed   Expected   Residual

        .00         37       31.30       5.70
       1.00         32       35.70      -3.70
       2.00         16       20.40      -4.40
       3.00          9        7.80       1.20
       4.00          4        2.80       1.20
                    --
    Total           98

    Chi-Square                 D.F.        Significance
      3.0694                    4              .5463
```

Data output 7.1

The data have been reduced to just five categories and each has an associated expected frequency. The residual is the difference between observed and expected. At the bottom is the actual chi-square statistic, the associated degrees of freedom (4 in this case as, once the categories had been clumped, there were five categories) and finally the significance value (*P* value).

Note: as we clumped several categories into one (four and above all became four) we should lose a further degree of freedom to give a total degrees of freedom of 3. The package does not account for clumping as it is done before the test is applied and therefore SPSS 'knows' nothing about it.

Chapter 7
The tests 1: tests
to look at
differences

Despite the fact that there are more high values and zero values than you would expect (indicative of a clumped distribution) the probability is well above 0.05. This indicates that the deviation from the Poisson expectations is nonsignificant and therefore we accept the null hypothesis that lice attack fish at random in this population.

MINITAB

MINITAB. Calculating a goodness of fit to a Poisson distribution is surprisingly awkward in MINITAB although not quite as awkward as in SPSS. However, it is a test that you will wish to carry out in many circumstances.

1 Make sure that the data are in a single column with the actual data, not the frequencies. In this case there will be 98 rows in the data set, one for each fish. Label the column 'No lice'.

2 Determine the mean number of lice per fish using the 'Stat' menu, then 'Basic statistics', then 'Descriptive statistics'. Move 'No lice' into the 'Variables' box. The output will confirm that there are 98 observations and give the mean as 1.143. (Or type 'Describe 'No lice'' at the MTB> prompt.)

The next steps show how to calculate the expected frequencies for a Poisson distribution in MINITAB. If you know how to do this by hand or in a spreadsheet you can skip directly to step 6.

3 First you should generate a table of the tallied observations: Go to the 'Stat' menu, then 'Tables' then 'Tally..'. Move 'No lice' into the 'Variables' box and make sure that the 'Counts' box is checked. Click 'OK'.

This output will appear in the 'Session' window:

```
Summary Statistics for Discrete Variables

No lice  Count
    0      37
    1      32
    2      16
    3       9
    4       2
    6       1
    7       1
   N=      98
```

Data output 7.2

Note that as there were no fish with five lice there is no count for 5 in the table. Now you can either cut and paste the two columns of figures from the 'Session' window into columns C2 and C3 of your MINITAB spreadsheet (remembering to click on the 'Use spaces as delimiters' box) or you can type the numbers in directly (Fig. 7.1). (Or type 'Tally c1;' at the MTB> prompt, followed by 'Store c2 c3.' at the SUBC> prompt. Remember to type a semicolon at the end of MTB> command if you want to bring up the SUBC> prompt.)

4 Now we need to generate expected numbers of lice for a Poisson distribution with the same mean as the sample. Go to the 'Calc' menu, then 'Probability distributions' then 'Poisson...'. In the dialogue box type the sample mean in the 'Mean:' box, type 'c2' in the 'Input column:' box and 'c4'

Chapter 7
The tests 1: tests
to look at
differences

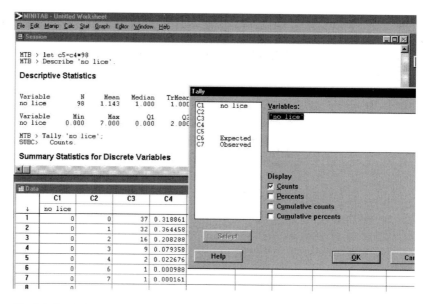

Fig. 7.1 Generating expected values using the 'Tally' command in MINITAB.

in the 'Optional storage:' box. If you don't select a storage column the output only goes to the 'Session' window. Click 'OK'. (Or type 'PDF c2 c4;' at the MTB> prompt, followed by 'Poisson 1.143.' at the SUBC> prompt. Replace 1.143 with the mean of your sample.)

5 To convert the probabilities generated into numbers you need to multiply by the total number of observations in the sample (98 in this sample). Go to the 'Calc' menu, then 'Mathematical expressions...'. In the dialogue box type c5 in the 'Variable (new or modified):' box. Then type 'c4 * 98' in the 'Expression:' box (replacing 98 with the number of observations in your sample). Click 'OK'. (Or type 'Let c5 = c4 * 98' at the MTB> prompt.)

6 Chi-square tests should not have expected frequencies that are less than one. In the example the expected frequencies for 6 and 7 are both less than one. Also, because there were no fish with five lice in the example there is no expected frequency for 5 in the column. In this example the best strategy is to pool all observations of 4 or more into a single observed and expected value. You can do this in several ways, either on a calculator, by hand or by using the 'Mathematical expression...' from the 'Calc' menu, selecting a row and typing the sum required. In this example selecting row 5 and typing '98 − 31.248 − 35.717 − 20.412 − 7.777' in the 'Expression box'. The numbers are the expected frequencies of 0, 1, 2 and 3 lice per fish respectively. (Or type 'Let c6(5) = 98 − 31.248 − 35.717 − 20.412 − 7.777' at the MTB> prompt. Replace c6 and (5) with the column and row you require and the numbers with those appropriate for your data.)

7 You must amalgamate the observed frequencies in exactly the same way as the expected. In this case there are a total of four observations of four or

Chapter 7
The tests 1: tests
to look at
differences

more lice per fish. You should now have one column of expected frequencies with no values less than one and one of observed. In the lice and fish example the following columns should result:

C6	C7
Expected	Observed
31.2484	37
35.7169	32
20.4122	16
7.7771	9
2.8460	4

8 Finally we reach the chi-square test itself. Amazingly there is no way to reach the required test using the menus. You will have to type at the command line (MTB>) the following: 'LET K1 = SUM((C7–C6)**2/C6)' (assuming that your expected values are in C6 and observed in C7) followed by 'PRINT K1' to see the result. For the example this will return the value of '3.059'.

9 Finally to determine whether this is significant or not (i.e. do we reject the null hypothesis that the lice are attacking the fish at random) we need to be sure on the degrees of freedom. In the example there were five possible values (so 4 d.f.) but one was an amalgamation so we loose a further d.f. (leaving 3 d.f.). You can either look up the value on a table in a statistics book or you can use the command line in MINITAB using the following commands:

```
MTB > cdf k1 k2;
SUBC> chisquare 3.   (replacing 3 with the d.f. required)

MTB > let k3=1–k2
MTB > print k3
```

This will give you:

```
Data Display
K3      0.382556
```

This value is well above the critical 0.05 level. Therefore, we accept the null hypothesis that lice attack fish at random in this population. If you look up the value in a statistics book you will see that the chi-square value required to reach the critical level of 0.05 for 3 d.f. is above 7.

Chapter 7
The tests 1: tests
to look at
differences

Excel

Excel. Most of the calculation steps are fairly straightforward and therefore ideal for a spreadsheet like Excel. Using the same example as for SPSS (above) I will assume that the data are available as a frequency table rather than a column containing all 98 observations. If it is not you can generate a frequency table from the raw data using the 'Frequency' command (or by hand).

1 To calculate Poisson expected values the only parameter you need to know is the mean. To find the mean of data in a frequency table you first need to find the product (what you get when you multiply the category value by the number of observations). If you have input the categories in column A with a title, number of observations in B with a title then go to cell C2 and type '=a2 * b2'. In this example this will be zero. Then find all the other products by dragging the little square in the bottom right corner of the cell down to the bottom of the list. Add up the number of observations in the B column by typing '=SUM(b2:b9)' in cell B11 and dragging this across to the C column to add up the products. Then divide the products by the number of observations to obtain the mean by typing '=c11/b11' in cell B13. Always use labels in other cells to make everything clear. The result will be something like that shown in Fig. 7.2.

Arial		11		**B**	*I*	U				

B13			=C11/B11			

	A	B	C	D	E	F
1	No lice	Count	lice * count			
2	0	37	0			
3	1	32	32			
4	2	16	32			
5	3	9	27			
6	4	2	8			
7	5	0	0			
8	6	1	6			
9	7	1	7			
10						
11	Sum	98	112			
12						
13	Mean	1.14286				
14						
15						

Fig. 7.2 Calculating the mean number of lice per fish from a table of frequencies in Excel. (*Note*: the numbers and letters around the sides of the data are the Excel row and column labels.)

Chapter 7
The tests 1: tests
to look at
differences

2 Calculation of the expected number of observations is surprisingly easy in Excel. Move to cell D2 (for convenience) and click on the 'Function wizard'. Select 'statistical' and then 'POISSON' from the very long list of options you are offered. A window with three lines to fill appears. In the first 'x' line use the cell number of the category, in this case A2. In the second type in the mean (1.143), don't use the cell number. Type a zero in the bottom line. This tells Excel that you do not want the cumulative probabilities. Then click on 'finish'. The probability of getting a zero in a Poisson distribution with a mean of 1.143 appears in the selected cell.

3 Step 2 calculated the probability of getting a particular number of lice. What is needed is the expected *number* of observations. To get this you must multiply the probability by the number of observations, 98 in this example. To do this make sure that cell D2 is selected and add a '*98' to the end of the formula giving: '=POISSON(A2,1.143,0)*98'. Press return and the number in the cell becomes the expected number of observations. Remember to replace 1.143 with your mean and 98 with the number of observations in your sample!

4 Select D2 again and drag its contents down the column (click on the little square in the bottom right corner of the cell and hold the mouse button as you move mouse to highlight all cells to D9 then let go). The column fills with the expected number of observations for each of the categories in column A.

5 Some of the expected values are less than one and need to be amalgamated. In this case it is best to have a category for lice numbers of 4 and above. Copy the observed data into column E but replace the number of times 4 lice were observed with the number of times 4 or more lice were observed. This is a total of four observations.

6 You have to amalgamate the expected frequencies too so that cell D6 holds not the expected number of fours but the expected number of 'four or mores'. There are several ways of achieving this but probably the easiest is to start with the 98 total and take off the expected number of 0, 1, 2 and 3. Type in cell D6 '=98−D2−D3−D4−D5'. This should give you an expected value of 2.845.

7 Chi-square uses the formula: '(observed − expected)^2/expected' for each category and then sums these to produce the final chi-square statistic. The formula is often quoted as $(O−E)^2/E$. In this example the observed values should be in column E and the expected in column D. So in cell F2 type in the formula: '=(E2−D2)^2/D2'. Once you have done that copy it down the F column to F6.

8 You should have five numbers in column F. These need to be added up to give the final chi-squared value. Type in any clear cell: '=sum(F2:F6)'. For convenience I used cell F8. You should get the value: 3.0599. This is the chi-square value that you would quote in a report.

9 Finally, what is the probability of getting this value of chi-square (or a higher value). Go to any clear cell. Click on the 'function wizard' and select 'CHIDIST' from the list. In the first box you input the chi-square value (or the cell that you used in step 8). In the second box you need to put the degrees of

Chapter 7
The tests 1: tests
to look at
differences

freedom. In this case there were five categories, but one of those was an amalgamation of many others. Therefore you should have three degrees of freedom (to be on the safe side). Input three and you should get a probability of 0.38. This value is well above 0.05 so you can infer that the distribution of lice on fish is not significantly different from random (Poisson distribution). We accept the null hypothesis that lice attack fish at random in this fish population.

Kolmogorov–Smirnov test

The Kolmogorov–Smirnov (K–S) test for goodness of fit has a variety of uses for large samples of continuous data. There are two main forms called the one-sample test or two-sample test. Both are used to compare two sets of data to determine whether they come from the same distribution. The one-sample version is more commonly used and compares experimental data with ex-pected distributions. The expected distribution may be derived from the data or may be completely independent of them. For example you may use the test to determine whether a set of tarsus length data differs from a normal distribution with the same mean and variance as the sample data before you use parametric analysis on it. The two-sample test can be used to compare a set of egg weight data from a population of ducks with a set from another site asking whether the distributions are the same.

Note: although the K–S tests appears similar to the *t*-test and the Mann–Whitney *U* test it is not aimed at the same question. K–S delivers a probability that two distributions are the same while the *t*-test is concerned with means and the M–W*U* test with medians. Two distributions may have identical means and medians and yet have differences elsewhere in their distributions.

An example

The weight in grams is recorded for a sample of 48 mice. This sample is part of an experiment and the researchers wish to know whether the weights are distributed normally before they go on to use parametric statistics.
Here is the data:

12.5	13.5	13.2	12.5	12.1	12.6	12.1	12.8
14.2	13.2	13.8	12.0	12.5	12.1	12.8	12.9
12.6	12.8	12.5	13.1	12.4	13.5	13.4	13.6

Continued

Chapter 7
The tests 1: tests
to look at
differences

13.0	14.1	12.6	13.2	13.8	13.8	13.9	14.0
14.1	12.1	12.9	14.5	13.2	14.1	12.5	12.5
15.0	12.6	13.0	13.5	14.0	12.9	12.4	12.8

SPSS

SPSS. This is a very simple test to access. Ensure all the data is in a single column. Select the 'Statistics' menu, then 'Nonparametric test', then '1-sample K-S...'. In the dialogue box move the name of the column you are testing into the 'Test Variable List:' box. Make sure that the 'Test distribution' has 'Normal' selected. Then click 'OK'. The following output will appear:

```
- - - - - Kolmogorov - Smirnov Goodness of Fit Test

   mouse_wt

   Test distribution  -  Normal                    Mean:  13.1083
                                        Standard Deviation:    .7178

            Cases:  48

            Most extreme differences
      Absolute        Positive        Negative        K-S Z      2-Tailed P
       .11475          .11475          -.08237         .7950        .5523
```

Data output 7.3

This output confirms the test and variable used, gives statistics: mean, standard deviation and number of observations. The last two lines of the output refer to the test itself. The only important bit is the '2-tailed P'. If this number is less than 0.05 then the distribution of the data is significantly different from normal. In this case the value is 0.5523 which is well above the critical 0.05 so we accept that the data is normally distributed.

MINITAB

MINITAB. Although there is no reference to the Kolmogorov–Smirnov test in the help file for MINITAB it is a very simple statistic to reach if you are testing a single column to see whether it follows a normal distribution or not although the Anderson–Darling test, considered below, is easier to reach.

Ensure all the data you wish to test is in a single column. Select the 'Graph' menu then 'Normal plot...'. In the dialogue box move the column you wish to test into the 'Variable:' box and select the Kolmogorov–Smirnov test. Then click 'OK'.

The output is a graph which shows a perfect normal distribution of data as a straight line and your data as a series of dots. The test compares the dots with the line. In the bottom left corner is the output from the statistical test. If the 'Approximate P-value' is less than 0.05 then the distribution is significantly different from normal. In this case the value is given as >0.15

Chapter 7
The tests 1: tests
to look at
differences

so it is not significantly different from normal. (Or at the MTB> prompt in the session window type '%NormPlot C1;' then at the SUBC> prompt type 'Kstest.')

Excel

Excel. There is no direct method using Excel.

Anderson–Darling test

One of many tests commonly encountered to determine whether a set of data follows a normal distribution or not. The P value reported is the probability of the data being normally distributed. If $P<0.05$ the data does not follow a normal distribution and parametric tests should not be used.

SPSS

SPSS. This test is not available in this package.

MINITAB

MINITAB. The Anderson–Darling test is part of the extensive 'Graphical' output for simple variable descriptions. The data should be in a single column. From the 'Stat' menu select 'Basic statistics' then 'Descriptive statistics…'. Move the name of the column containing the data into the 'Variables:' box. Select the 'Graphical form' option. Click 'OK'.

An output containing several graphs and many descriptive statistics appears. The first part of the numerical output give the statistics for the Anderson–Darling test ('A-squared') followed by the P value associated with the statistic. If the P value is less than 0.05 the data is significantly different from a normal distribution and it is inadvisable to use parametric statistics. (See Fig. 6.7)

Excel

Excel. The Anderson–Darling test is not available in this package.

Do the observations from two groups differ?

The two groups can be paired, repeated or related samples or they can be independent. Paired measures are considered first.

Paired data (a.k.a. related or matched data)

Paired samples or paired comparisons occur when a single individual is tested twice (e.g. before and after) or sampling station retested. Another possible use occurs when an individual is, or individuals of a clone are, divided and then subjected to two treatments. Three tests are considered below: the paired *t*-test; Wilcoxon's signed ranks test (page 75) and the sign test (page 77).

Chapter 7
The tests 1: tests
to look at
differences

Paired t-test

The data must be continuous and, at least approximately, normally distributed. The variances of the two sets must be homogeneous (this can be tested by the Levene test). The null hypothesis is that there is no difference between the two columns and they could come from the same data set.

An example

It is suggested that the building of a power station will affect the amount of particulate matter in the air. However, there are only three readings available for the month before the project got underway. The sites where the three readings were taken were revisited once the station was complete.

Case (site)	Before	After
1	34.6	41.3
2	38.2	39.6
3	37.6	41.0

SPSS

SPSS. Arrange the data into two columns of equal length such that each row represents one individual. The columns should be labelled 'before' and 'after' as this will make it easier to interpret the output.

Under the 'Statistics' menu choose 'Compare means' and then 'Paired-samples *t* test'. In the dialogue box that appears highlight both 'before' and 'after' as your paired variables then click 'OK'.

The output will come in two parts. The first gives some information about the data like this:

```
t-tests for Paired Samples

                  Number of           2-tail
Variable            pairs     Corr     Sig        Mean        SD       SE of Mean
-----------------------------------------------------------------------------------
AFTER                                             40.6333     .907       .524
                     3        -.749    .462
BEFORE                                            36.8000    1.929      1.114
-----------------------------------------------------------------------------------
```

Data output 7.4

It tells you the names of the two variables, how many pairs there were. The next two columns refer to a Pearson product–moment correlation done on the data with the value of 'r' and the probability that it differs from 0. Next comes summary information about the data, the mean, standard deviation and

[73]

Chapter 7
The tests 1: tests
to look at
differences

standard error of the two sets of data. The second part of the output is the
report from the *t*-test itself:

```
Paired Differences
   Mean        SD     SE of Mean  |       t-value        df       2-tail Sig
---------------------------------- |-------------------------------------------
  3.8333      2.676      1.545     |        2.48           2          .131
95% CI (-2.815, 10.482)
```

Data output 7.5

The first three columns refer to the mean difference between pairs of data,
followed by their standard deviation and standard error. Under this comes
the confidence interval of the mean saying it is 95% likely to be between
the two values given. The 95% CI is very wide because there are only three
pairs of data. Finally comes the *t*-test itself with the result of the test (the
value given can be looked up on a Student's *t*-table). Then the degrees of
freedom (number of pairs minus one and the probability of this *t* value (or
larger) occurring if the null hypothesis is correct). In this case the probability
is 0.131 (or 13.1%) and there is no significant difference between the two
columns. However, with such a small set of data getting a significant result is
extremely unlikely.

MINITAB

MINITAB. First input the data into two columns one for before the other after.
There is no need to have a separate column to label the individual sites
although this might help you interpret the results.
1   You need to create a new column that contains the difference between
the first observation and the second. Go to the 'Calc', select 'Mathematical
expressions...'. Type 'diff' in the 'Variable (new or modified):' and 'Before' —
'After' (either by typing or double-clicking) in the 'Expression:' box.
2   The null hypothesis is that the differences are not significantly different
from zero. This is tested using a one-sample *t*-test. Go to the 'Stat' menu, then
'Basic statistics', then '1-sample t...'. In the dialogue box put 'diff' in the
'Variables:' box. Select the 'Test mean' option and make sure that the 'not
equal' option is selected from the pull-down menu. Click 'OK'.
You get the following output from the example.

```
T-Test of the Mean

Test of mu = 0.00 vs mu not = 0.00

Variable     N     Mean     StDev    SE Mean      T      P-Value
diff         3    -3.83     2.68      1.55      -2.48     0.13
```

Data output 7.6

3   The output confirms that you are testing the mean of the variable 'diff'

Chapter 7
The tests 1: tests
to look at
differences

against a hypothesized mean of 0. There are three pairs of data in the example giving a value of 3 for N. There are then summary statistics for 'diff'. Finally a T value and then the important value is the 'P Value'. If this is less than 0.05 then you must reject the null hypothesis that the difference between before and after is zero. In this case the value of 0.13 indicates that there is not enough evidence in the three pairs of data to reject the null hypothesis. However, with so little data in the example it is not surprising that you can't detect a significant effect.

Excel

Excel. The data should be input in two columns of equal length. If the spreadsheet were set up using the example exactly as the table above, then the data for 'before' would be in cells b2, b3 and b4, while that for 'after' would be in cells c2, c3 and c4. In Excel it doesn't really matter where the data is on the spreadsheet as long as you know the cell numbers.

Select an empty cell where you want the result reported. Then there are two ways of achieving the same result.

1   Go to the 'Function Wizard', select 'statistical' and then 't-test'. Define the first array as 'b2:b4' and the second as 'c2:c4'. Select 'tails' as '2' (you will nearly always require a two-tailed test) and the 'type' as '1' (this selects a paired test in Excel). The probability or P value of '0.131' will then appear in the cell.

2   Type in 'TTEST' followed by the first and last cell of the first column separated by a colon, then the same for the second column. Then the number of tails in the test (usually 2) and then a 1 to ask for a paired test. In this case you would type 'TTEST(b2:b4,c2:c4,2,1)' and the probability will appear in the cell.

Wilcoxon's signed ranks test

This test is the nonparametric equivalent of the paired t-test. It has far fewer assumptions about the shape of the data although it does assume that the data are on a continuous scale of measurement. This means that any type of length, weight etc. will be suitable. The test is somewhat less powerful than the paired t-test. A minimum of six pairs of data is required before the test can be carried out.

An example

In this example the 'individuals' are sampling stations in a river system and the data are measures of flow. The investigator wishes to know if the flow is significantly different on the two days. The null hypothesis is that there is no difference in flow.

Chapter 7
The tests 1: tests
to look at
differences

Station	Day 1	Day 2
1	268	236
2	260	241
3	243	239
4	290	285
5	294	282
6	270	273
7	268	258

SPSS

SPSS. Arrange the data into two columns of equal length such that each row represents one individual. The columns should be labelled.

From the 'Statistics' menu choose 'non-parametrics' and then '2-related samples'. As a default the 'Wilcoxon' test should be checked but ensure this is the case. Select the two variables 'day_1' and 'day_2' in this example and click 'OK'.

The output will look like this:

```
- - - - - Wilcoxon Matched-Pairs Signed-Ranks Test

     DAY_1
with DAY_2

   Mean Rank    Cases

       4.50        6   - Ranks  (DAY_2 LT DAY_1)
       1.00        1   + Ranks  (DAY_2 GT DAY_1)
                   0     Ties   (DAY_2 EQ DAY_1)
                   -
                   7     Total

        Z =   -2.1974              2-Tailed P =   .0280
```

Data output 7.7

The test classifies the paired data into three categories: those where 'day_2' is less than 'day_1'; those where it is greater and those where they are the same (ties). In this case there are six of the first, one of the second and no ties. It also ranks the differences (from smallest difference as rank one to the largest as rank seven). In this case the one pair where the flow is greater on 'day_2' is also the smallest difference as it is ranked '1'. The output value of 'Z' at the bottom may be looked up in a table but the result is given anyway as a '2-tailed P'. In this case $P < 0.05$ so the null hypothesis must be rejected. The alternative hypothesis that the flows were different on the two days must be accepted.

MINITAB

MINITAB. This test is not achievable in a single step. However if the data are arranged in two columns you can carry out an analogous test to the paired *t*-test.

Chapter 7
The tests 1: tests
to look at
differences

1   You need to create a new column that contains the difference between the first observation and the second. Go to the 'Calc', select 'Mathematical expressions…'. Type 'diff' in the 'Variable (new or modified):' and 'Day 1'-'Day 2' (either by typing or double-clicking) in the 'Expression:' box.

2   The null hypothesis is that the median of the differences is not significantly different from zero. This is tested using a one-sample Wilcoxon test. Go to the 'Stat' menu, then 'Nonparametrics', then '1-sample Wilcoxon…'. In the dialogue box put 'diff' in the 'Variables:' box. Select the 'Test median' option, leave the value as 0.0 and make sure that the 'not equal' option is selected from the pull-down menu. Click 'OK'.

You get the following output from the example.

```
Wilcoxon Signed Rank Test

TEST OF MEDIAN = 0.000000 VERSUS MEDIAN N.E. 0.000000

                 N FOR    WILCOXON             ESTIMATED
            N    TEST    STATISTIC   P-VALUE    MEDIAN
diff        7     7         27.0     0.035       10.50
```

Data output 7.8

3   The output confirms that you are testing the null hypothesis that the median of 'diff' (the difference between the flows on the two days) is zero. The important value is the 'P-value' which is 0.035 in this example. This is less than the critical 0.05 and the null hypothesis should be rejected. The alternative hypothesis H_1 is that the median of the differences is not equal to zero. This means that the flows on day 1 and day 2 were different. Inspection of the raw data or the estimated median shows that flow was greater on day 1.

Excel

Excel. There is no easy method of doing this test with Excel.

Sign test

This is a very simple test that makes almost no assumptions about the form of the data only that it is possible to compare them in some way to decide which is larger. The test is of very low power but is very safe! (i.e. it is a conservative test and type I errors are very unlikely). The sign test should only be used when there are large numbers of paired observations.

The test works using the assumption that if two sets of observations are not different then there will be the same number of pairs when A is bigger than B as there are when B is bigger than A. Therefore the actual values of the data points are relatively unimportant as long as they can be compared to see which is the larger. This means the test is not very sensitive to poor quality of data.

Chapter 7
The tests 1: tests
to look at
differences

Example

The same data as was used as for the Wilcoxon test.

SPSS

SPSS. Arrange the data into two columns of equal length such that each row represents one individual. The columns should be labelled appropriately.

From the 'Statistics' menu choose 'non-parametrics' and then '2-related samples'. As a default the 'Wilcoxon' test should be checked, uncheck this and check the 'sign test' instead. Select the two variables 'day_1' and 'day_2' in this example and click 'OK'.

The output (using the example data described earlier for the 'Wilcoxon Signed Ranks Test') will look like this:

```
DAY_1 with DAY_2

    Cases
       6   - Diffs (DAY_2 LT DAY_1)
       1   + Diffs (DAY_2 GT DAY_1)     (Binomial)
       0     Ties                       2-Tailed P =      .1250
       -
       7     Total
```

Data output 7.9

For the seven pairs of observations there are six where 'day_2' is greater than 'day_1'. The probability of this or a more extreme result occurring by chance is 0.1250 or 12.5% and this is given as the '2-tailed P'. In fact it is giving the probability of getting 0, 1, 6 or 7 out of 7 the same direction given that we are expecting the chance of getting a higher or lower value to be 0.5 (i.e. equal chance of A bigger than B as for B bigger than A).

In this case we would not reject the null hypothesis that the two days had different flows. This shows clearly that the 'Sign test' is of very much lower power than the Wilcoxon Signed Ranks Test. In fact the sign test is really only useful as the number of paired observations becomes quite large.

MINITAB

MINITAB. This test is carried out in a very similar way to the Wilcoxon's signed ranks test. Arrange the data in two columns. Label the columns appropriately.

1 Create a new column that contains the difference between the first observation and the second. Go to the 'Calc', select 'Mathematical expressions…'. Type 'diff' in the 'Variable (new or modified):' box and 'Day 1' – 'Day 2' (either by typing or double-clicking) in the 'Expression:' box. Obviously you will have to replace these names with the appropriate names for the data columns.

2 The null hypothesis is that there are equal numbers of positive and negative differences. This is tested using a one-sample sign test. Go to the 'Stat' menu,

[78]

Chapter 7
The tests 1: tests
to look at
differences

then 'Nonparametrics', then '1-sample Sign…'. In the dialogue box put 'diff' in the 'Variables:' box. Select the 'Test median' option, leave the value as 0.0 and make sure that the 'not equal' option is selected from the pull-down menu. Click 'OK'.

You get the following output from the example:

```
Sign test for median

Sign test of median = 0.00000 versus  N.E.   0.00000

                N   BELOW  EQUAL  ABOVE   P-VALUE      MEDIAN
diff            7     1      0      6     0.1250       10.00
```

Data output 7.10

3   The output confirms that you are testing the null hypothesis that the median of 'diff' (the difference between the flows on the two days) is zero. The important value is the 'P-value' which is 0.1250 in this example (i.e. there is a 12.5% chance of getting this result or a more extreme one). This is greater than the critical level of $P = 0.05$ and the null hypothesis should be accepted. This is the same data as was used in the Wilcoxon's signed ranks test that rejected the null hypothesis, thus demonstrating what a conservative test it is.

Excel. There is no direct method. However, if you compare each pair of data and count up cases where the first data set is greater then you can use the 'binomdist' function in Excel to work out the P-value. The binomial distribution is described in Chapter 5.

Using the example above, there are seven pairs of data. In six of the seven the first column is greater. We are assuming a null hypothesis that there is no difference between the two columns. If that is the case then we expect an equal number of observations where the first column is greater and when the first is smaller.

1   In Excel the probability of getting 6 plusses out of 7 can be determined fairly easily. After selecting an empty cell where you want the result to be displayed there are two methods. Either: Use the 'Function Wizard', select 'Binomdist', click 'next' and you will be confronted with four boxes. In the first type the number of 'successes', in this case 6. In the second the number of 'trials', in this case 7. Next the probability of getting a 'success', in this case we assume 0.5 (we assume an equal chance of getting a plus or a minus). In the last box type 0 to indicate that you don't require the cumulative probability. This will give you the result 0.054. Or: type in the commands described above directly: '=BINOMDIST(6,7,0.5,0)'. This is not the complete answer. It only tells you the chance of getting 6 out 7. What you need from the sign test is the chance of getting a result that extreme or more. For seven trials that means 6 or 7 as well as 0 or 1 because the test is two-tailed.

2   In a new empty cell for each possibility, repeat the previous commands but

Chapter 7
The tests 1: tests
to look at
differences

replacing the 6 successes with 7, 0 and 1. A good way to do this is to put these numbers into a column then use the cell number instead of typing in the number. For example, if cell A1 has a 7 in it, A2 has 6 etc. then just type into cell B1 '=BINOMDIST(A1,7,0.5,0)' and then copy this down the next four cells by clicking on the small black square in the bottom right corner of the cell and dragging it down to cover the cells required.

3   Total up the four probabilities by typing '=SUM(B1:B4)' in an empty cell. This should give you the answer 0.125. This can be interpreted as a 12.5% chance of getting a result as extreme as 6 out of 7 or more. We have to accept the null hypothesis that there is nothing happening between the two sampling events because the value is well above $p = 0.05$. This shows the low power of the test. In fact with a sample size this small you need to get 7 out of 7 or 0 out of 7 to get a significant result.

Unpaired data

Unpaired samples or unpaired comparisons occur when a single individual is measured or tested only once. There will, therefore be two totally separate groups of observations making up the two samples. Two groups are often obvious: e.g. males and females; kudu and eland. However, the distinction between the groups may be rather arbitrary: e.g. eastern and western; large and small. Three tests are considered below: the independent samples *t*-test; a one-way analysis of variance (page 85) and the Mann–Whitney *U*-test (page 92).

t-test

The independent samples *t*-test is the more usual form of the *t*-test and if the term '*t*-test' is not qualified then this is what is being referred to. The null hypothesis is that the two sets of data are the same. (Actually the null hypothesis is that the two sample means come from a population with the same true mean, μ.) The *t*-test assumes that the data are continuous, at least approximately normally distributed and that the variances of the two sets are homogeneous (the same!). If possible, these assumptions should be tested before the test is carried out, although test are often incorporated into the *t*-test in statistical packages. If the two sets of observations do not have the same variance then there are ways to adjust the result of the *t*-test to compensate. It is these adjustments that are often incorporated into the test in the statistical package.

An example

The weights of five grains have been measured from each of two experimental cultivars called '*premier*' and '*super*'. Each grain has been weighed to the nearest 0.1 mg. The researcher wishes to determine whether the grain weight

[80]

Chapter 7
The tests 1: tests
to look at
differences

is the same in the two cultivars. The null hypothesis (H_0) is that the two cultivars have the same grain size. The alternative hypothesis (H_1) is that the two cultivars have different mean grain size.

Premier	Super
24.5	26.4
23.4	27.0
22.1	25.2
25.3	25.8
23.4	27.1

SPSS

SPSS. Input all of the observed data into a single column. Use another column for the labelling of the groups. This may seem wasteful but it is a much easier system when it comes to multiway analysis where each item of data will belong to several groups simultaneously.

In the example below five grains from each of two cultivars of crop plant have been placed in the second column. The two cultivars have been coded as '1' and '2' in the first column. The package only allows restricted labels so the ideal label 'Grain size (mg)' has been shortened to 'grain_sz'.

cultivar	grain_sz
1	24.5
1	23.4
1	22.1
1	25.3
1	23.4
2	26.4
2	27.0
2	25.2
2	25.8
2	27.1

Under the 'Statistics' menu choose 'Compare means' and then 'Independent-Samples T Test'. In the dialogue box that appears move 'grain_sz' into the 'Test Variable(s):' box by first highlighting it and then clicking on the

Chapter 7
The tests 1: tests
to look at
differences

appropriate move button. Next move 'cultivar' into the 'Grouping Variable' box. It will appear as 'cultivar(? ?)'. You need to click the 'Define Groups...' button and then input '1' in 'Group 1:' and '2' in 'Group 2:' before clicking 'Continue'. This will return you to the first dialogue box. You will see that 'cultivar(? ?)' is now 'cultivar(1 2)'. Click 'OK' to run the test.

The output will come in two parts. The first gives some information about the data like this:

```
t-tests for Independent Samples of CULTIVAR

                            Number
Variable                    of Cases        Mean        SD    SE of Mean
-----------------------------------------------------------------------
GRAIN_SZ

CULTIVAR 1                      5          23.7400      1.218      .545
CULTIVAR 2                      5          26.3000       .806      .361
-----------------------------------------------------------------------

        Mean Difference = -2.5600

        Levene's Test for Equality of Variances: F= .762   P= .408
```

Data output 7.11

This confirms the type of test. Then the table shows that the variable used was 'grain_sz' and that there were two groups of 'cultivar' called '1' and '2' each with 5 cases (observations). Then come some simple descriptive statistics of the two samples: their mean; 'SD' (standard deviation) and 'SE' (standard error of the mean). Under the table is the difference between the two sample means. (Note that it is given as a minus figure as the mean of cultivar 2 was bigger than that for cultivar 1. If the cultivars had been labelled the other way round this would not be a negative number.)

Finally comes the line for the 'Levene's test for equality of variances'. As the t-test assumes that the two samples have equal variance SPSS sensibly tests this every time you carry out a t-test. The important figure is the last one. In this case it is '$P=0.408$' so there is no evidence that the variances are unequal. If the P value is lower than 0.05 then you should be wary about using the t-test and should use the Mann–Whitney U test instead.

The second part of the output is the result of the t-test itself:

```
t-test for Equality of Means
                                                                   95%
Variances   t-value      df    2-Tail Sig    SE of Diff        CI for Diff
-------------------------------------------------------------------------
Equal        -3.92        8       .004          .653        (-4.066, -1.054)
Unequal      -3.92      6.94      .006          .653        (-4.107, -1.013)
-------------------------------------------------------------------------
```

Data output 7.12

The output appears on two lines. The upper one, 'equal', is the standard t-test and the lower one 'unequal' is a more conservative version that compensates for the possible problems caused by difference in variances by using

Chapter 7
The tests 1: tests
to look at
differences

a reduced value for the degrees of freedom in the test. The lower the *P* value in the Levene test the bigger the difference between the two lines. In most cases you should not need to worry about the fact that there are two versions of the test because in most cases, as in this example, they will tell the same story.

The '*t*-value' is the actual result of the test that can be looked up on a Student's *t*-table. 'df' is the degrees of freedom of the test. This will be two less than the total number of observations in the two samples. Next comes the important bit, labelled '2-Tail Sig' it is the probability that the null hypothesis is correct. This is what is usually called the *P* value. In this case the *P* value is much less than 0.05 so it is clear that the null hypothesis is extremely unlikely to be true. In fact there is only a 0.4% chance that the null hypothesis is true using the basic *t*-test, 0.6% using the more conservative version.

Finally comes a '95% CI for Diff' column. This gives the range of difference between the two means within which 95% of samples are likely to come. It shows that cultivar 2, '*Super*', is always going to be heavier than cultivar 1, '*Premier*'.

MINITAB

MINITAB. There are two methods. In the first the two groups should be in two columns, in the second all the data is in one column and another column is used to identify the groups. The second method appears wasteful but it is a required strategy in more complex analysis.

Method 1. Input the data into two columns and label the columns appropriately. From the 'Stat' menu select 'Basic statistics' and then '2-sample t...'. In the dialogue box that appears select the 'Samples in different columns' option. Then place the appropriate column labels ('*Premier*' and '*Super*' in the example); in the two boxes labelled 'First:' and 'Second:'.

Make sure the pull down menu reads 'not equal' and the 'Confidence level:' is set at 0.95 (this means that the mean for the difference between the two means will be calculated with a 95% confidence interval). Leave the 'Assume equal variances' box unchecked. Click 'OK' to run the test. (Or type 'twosample c1 c2' at the MTB> prompt in the session window.)

The following output appears:

```
Two Sample T-Test and Confidence Interval

Twosample T for Premier vs Super
             N      Mean     StDev    SE Mean
Premier   5      23.74     1.22      0.54
Super     5      26.300    0.806     0.36

95% C.I. for mu Premier - mu Super: ( -4.16,  -0.96)
T-Test mu Premier = mu Super (vs not =): T= -3.92  P=0.0078  DF=  6
```

Data output 7.13

This output confirms the test was a *t*-test and confirms the names of the two variables. It then gives summary statistics for the two groups: number of

Chapter 7
The tests 1: tests
to look at
differences

observations ('N'), mean, standard deviation and standard error of the mean. The next line gives the 95% confidence interval for the mean difference between the two groups. The references to 'mu' are to the Greek letter, μ, which is used to denote a mean. In the last line is the output for the *t*-test itself, confirming that a test of equal means is being made. The value of *T* is given as −3.92 and the important *P* value as 0.0078. This value is much less than 0.05 so we reject the null hypothesis and accept the H$_1$ that the two groups have different means. Inspection of the summary statistics shows that '*Super*' has larger grain size.

The degrees of freedom ('DF') is given as 6. There should be 8 degrees of freedom for the example but as the assumption of equality of variance was not made a correction is applied to the degrees of freedom to make the test more conservative. If the 'Assume equal variances' is checked then the example data will give 8 degrees of freedom.

Method 2. Input all the data into a single column and use a second column to label the data. These labels should be integers. In the example it is probably best to label '*Premier*' as '1' and '*Super*' as '2'. Label the columns appropriately. In the example the data column should be labelled 'Grain sz' and the group codes as 'Cultivar'.

From the 'Stat' menu select 'Basic statistics' and then '2-sample t...'. In the dialogue box that appears select the 'Samples in one column' option. Then place the column with the data in the 'Samples:' box and the one with the group codes in the 'Subscripts:' box. Make sure the pull down menu reads 'not equal' and the 'Confidence level:' is set at 0.95. Leave 'Assume equal variances' unchecked. Click 'OK' to run the test. (Or type 'twot c1 c2' (assuming data is in c1) at the MTB> prompt in the session window.)
You get the following output:

```
Two Sample T-Test and Confidence Interval

Twosample T for Grain sz
Cultivar  N      Mean     StDev    SE Mean
1         5      23.74    1.22     0.54
2         5      26.300   0.806    0.36

95% C.I. for mu 1 - mu 2: ( -4.16,  -0.96)
T-Test mu 1 = mu 2 (vs not =): T= -3.92   P=0.0078   DF=   6
```

Data output 7.14

Apart from the code names '1' and '2' replacing the group names the output is identical to that produced in Method 1.

Excel

Excel. In this case the data may be anywhere on the spreadsheet. As long as you know the cell locations of the two groups there is no problem. However, in practice, it is much easier if the data is either input exactly as in the SPSS example above, with one column defining the group and another containing

Chapter 7
The tests 1: tests
to look at
differences

the actual data or in two adjacent, and clearly labelled, columns. Assuming you have input the data in an identical format to the SPSS example for the *t*-test then the data for the first cultivar is in cells b2 to b6 and for the second cultivar in cells b7 to b11. There are now two methods that may be used.

Method 1. Go to the 'Function Wizard', select 'STATISTICAL' and then 'TTEST'. Define the first array as 'b2:b6' and the second as 'b7:b11'. These arrays may be defined by selecting the box then clicking and dragging over the appropriate cells on the spreadsheet. Select 'tails' as '2' (you will nearly always require a two-tailed test) and the 'type' as '2' (this selects a standard *t*-test). The probability or *P* value of '0.0044' will then appear in the cell.

Method 2. Type in '=TTEST' followed by the first and last cell of the first column separated by a colon, then the same for the second column. Then the number of tails in the test (usually 2) and then a 2 to ask for a standard *t*-test. In this case you would type '=TTEST(B2:B6,B7:B11,2,2)' and the probability will appear in the cell. Excel does not report the *t*-value.

If you are concerned that the variances of the two samples may not be equal, or you know it to be the case, then you should not use the standard *t*-test. Excel allows you to carry out a *t*-test that does not assume homogeneity of variances. This is easily accessed by using a type 3 test instead of a type 2. If you use a type 3 *t*-test with this example the probability should be reported as '0.0058'.

One-way ANOVA

Using analysis of variance (ANOVA) to determine whether just two groups have the same mean may seem like overkill. This may be the simplest use of ANOVA but it still works and gives the same answer as the *t*-test. I am of the opinion that the fact that the *t*-test is restricted to two groups makes the use of ANOVA preferable in this situation because you don't have to learn a new test when you consider more than two groups.

ANOVA has the same basic assumptions as the *t*-test: that the data are continuous; at least approximately normally distributed and that the variances of the data sets are homogeneous. These assumptions should be tested before the test is carried out. The null hypothesis is that the sets of data have the same mean. (The way that ANOVA actually approaches this is to have a null hypothesis that the variation within groups is the same as variation between groups.)

An example

In the illustrations of the use of the test for the packages I will be using the same example data set as in the *t*-test. With two samples of five observations each and each sample coming from a different cultivar of a crop plant.

Chapter 7
The tests 1: tests
to look at
differences

The ideal ANOVA table that you would include in a write-up or publication should appear as something along the lines of the table shown here.

Source	d.f.	SS	MS	F	P
Cultivar	1	16.38	16.38	15.36	0.004**
Residual	8	8.53	1.07		
Total	9	24.92	2.77		

d.f. is the degrees of freedom, SS is the sum of squares, MS is the mean square (SS/d.f.), F is the ratio of within group variation to between group variation (in this case it is the MS for cultivar/residual MS). The P value is the important one in this case; it is less than 0.05 so we reject the null hypothesis that the two cultivars have the same mean grain size. The two asterisks indicate a highly significant result and are often used when $P < 0.01$. Compare this table with the output obtained using the packages below to see where the various numbers have come from. (Note the word 'residual' often appears as 'error'.)

If you wish to write this result in the text of a report the standard way would be as follows: analysis of variance showed that the grain size of the two cultivars was significantly different ($F_{1,8} = 15.36$, $P < 0.01$). The two subscripted numbers after the F indicate the degrees of freedom for the between group and within group variance.

SPSS

SPSS. As with the *t*-test all the data are placed into a single column and another column is used for the labelling of the groups. This may seem wasteful but it is a much easier system when it comes to more complicated analyses.

There are at least three routes to this test in SPSS. Unfortunately they give rather different outputs. I will describe them in detail below. I suggest you try all methods as the comparisons may help you understand how ANOVA tables work, and especially which parts of the tables to look at.

Method 1. Under the 'statistics' menu choose 'compare means' and then 'One-way ANOVA'. In the dialogue box that appears move 'grain_sz' into the 'Dependent list:' box by first highlighting it and then clicking on the appropriate move button. Next move 'cultivar'into the 'Factor' box. It will appear as 'cultivar(? ?)'. You need to click the 'Define Range...' button and then input '1' in 'Minimum' and '2' in 'Maximum' before clicking 'Continue'. This will return you to the first dialogue box (Fig. 7.3). You will see that 'cultivar(? ?)' is now 'cultivar(1 2)'. Click 'OK' to run the test.

Chapter 7
The tests 1: tests
to look at
differences

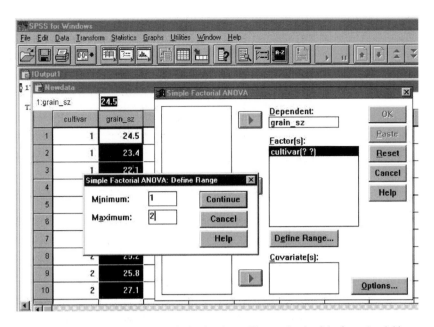

Fig. 7.3 Using SPSS for one-way analysis of variance. The two levels of the factor 'strain' have already been defined as '1' and '2'.

This is the output you will get:

```
- - - - - O N E W A Y - - - - -

       Variable  GRAIN_SZ
    By Variable  CULTIVAR

                              Analysis of Variance

                            Sum of        Mean          F      F
          Source      D.F.  Squares      Squares      Ratio   Prob.

Between Groups          1   16.3840      16.3840     15.3624  .0044
Within Groups           8    8.5320       1.0665
Total                   9   24.9160
```

Data output 7.15

This confirms that the data for 'grain_sz' have been grouped using 'cultivar' and then goes on to a standard ANOVA table. The top line of the table is the important one; this is the variation between groups (i.e. between the two cultivars in this example). The second line is the variation within the groups that is being used as a comparison.

The first column is 'D.F.' or degrees of freedom. As there are two groups there is one degree of freedom between groups. There were five samples in each group giving two lots of four degrees of freedom, and therefore eight in total. Next come some calculation columns: sum of squares (often called SS) and mean squares (MS). These are customarily included in ANOVA tables. The mean square value is the sum of square value divided by the degrees of freedom.

Chapter 7
The tests 1: tests
to look at
differences

Finally comes the important bit; the *F* ratio. This is the mean square for between groups divided by that for within groups. If there is the same amount of variation between and within groups this will give an *F* ratio of exactly one. In this case the *F* ratio is over 15. This value could be looked up on an *F* table, although SPSS gives you the *P* value anyway, labelling it as *F* prob. The value is 0.0044 indicating that the two groups are highly significantly different, as $P < 0.01$.

'Options…': one of the assumptions of ANOVA is that variances are equal. In the *t*-test in SPSS this is tested automatically. In ANOVA it is available under the 'Options…' button. Just check the box labelled 'Homogeneity of Variance' and click 'OK' before running the test. If you do this a little extra output appears after the ANOVA table.

```
Levene Test for Homogeneity of Variances

   Statistic    df1   df2      2-tail Sig.
      .7616       1     8          .408
```

Data output 7.16

The important value is given the label '2-tail Sig.'. The critical value is usually 0.05 and if the value given here is less than that ANOVA should not be used but a Mann–Whitney *U* test used instead.

Method 2. Under the 'statistics' menu choose 'ANOVA Models' and then 'Simple Factorial…'. In the dialogue box that appears move 'grain_sz' into the 'Dependent' box by first highlighting it and then clicking on the appropriate move button. Next move 'cultivar' into the 'Factor(s)' box. It will appear as 'cultivar(? ?)'. You need to click the 'Define Range…' button and then input '1' in 'Minimum' and '2' in 'Maximum' before clicking 'Continue'. This will return you to the first dialogue box. You will see that 'cultivar(? ?)' is now 'cultivar(1 2)'. Click 'OK' to run the test.

(Note that unlike method 1 there is no option to run a Levene test for homogeneity of variances at the same time as running the ANOVA. For this reason you may prefer to use method 1. However, more complicated ANOVA tests are not possible under the 'Compare Means' menu so it may be better to become familiar with this method.)

As in the first method this confirms that the data for 'grain_sz' has been grouped using 'cultivar' and then goes on to a rather less standard ANOVA table. Remember that this output is designed to cope with many factors and therefore extra lines that appear totally superfluous here are useful in more complex experiments. The important line is repeated three times, although you should read along the line labelled 'cultivar'. This is the variation that was labelled 'between groups' in the first method. The line labelled 'residual' corresponds to the one labelled as 'within groups'.

The first column is 'sum of squares', then 'DF' or degrees of freedom, then 'mean square' (the mean square value is the sum of square value divided

Chapter 7
The tests 1: tests
to look at
differences

```
* * *   A N A L Y S I S   O F   V A R I A N C E   * * *

          GRAIN_SZ
     by   CULTIVAR

          UNIQUE sums of squares
          All effects entered simultaneously

                                 Sum of                Mean              Sig
Source of Variation              Squares      DF       Square      F     of F

Main Effects                     16.384       1        16.384    15.362  .004
   CULTIVAR                      16.384       1        16.384    15.362  .004

Explained                        16.384       1        16.384    15.362  .004

Residual                          8.532       8         1.066

Total                            24.916       9         2.768


10 cases were processed.
0 cases (.0 pct) were missing.
```

Data output 7.17

by the degrees of freedom). As there are two cultivars they have one degree of freedom. There were five samples within each cultivar giving two lots of four degrees of freedom and therefore eight in total for 'residual'.

Finally comes the important bit; the F ratio, labelled 'F' here. This is the mean square for 'cultivar' divided by that for 'residual'. The value of F is 15.362. SPSS gives you the P value associated with this value of F and these degrees of freedom and labels it 'Sig of F'. In biology we usually look for a value less than 0.05. Here the probability is 0.004 and indicates that the mean grain size of the two cultivars are significantly different. The output of this method is certainly not appropriate for inclusion in a write up as it stands, create a table in the form I gave in the example.

Method 3. A third method in SPSS is very similar to method 2, except that you select 'General Factorial...' instead of 'Simple Factorial...'. Follow the same instructions as for method 2 and you get output as follows:

```
* * * * * * A n a l y s i s   o f   V a r i a n c e -- design  1 * * * * * *

Tests of Significance for GRAIN_SZ using UNIQUE sums of squares
Source of Variation           SS        DF        MS         F   Sig of F

WITHIN+RESIDUAL              8.53        8       1.07
CULTIVAR                    16.38        1      16.38      15.36    .004

(Model)                     16.38        1      16.38      15.36    .004
(Total)                     24.92        9       2.77

R-Squared =          .658
Adjusted R-Squared = .615

- - - - - - - - - - - - - - - - - - - - - - - - - - - - - - - - - - - - -
```

Data output 7.18

As with method 2 you are most interested in the line labelled 'cultivar' and the P value, labelled as 'Sig of F' at the end of the line. The rest of the table

Chapter 7
The tests 1: tests
to look at
differences

MINITAB

is similar to that from method 2. The last two lines are part of a regression analysis and should be ignored here.

MINITAB. As with the *t*-test there are two ways of inputting the data. In the first the two groups should be in two columns, in the second all the data is in one column and another column is used to identify the groups. The second method appears wasteful but it is required in more complex analyses.

Method 1. Input the data from the two groups into two separate columns and label appropriately. From the 'Stat' menu select 'ANOVA' then 'Oneway (unstacked)'. In the dialogue box ensure that both columns are in the 'Responses (in separate columns):' box. Click 'OK'. (Or type 'aovo c1 c2' at the MTB> prompt in the session window.)

You get this output:

```
One-Way Analysis of Variance

Analysis of Variance
Source       DF        SS        MS        F        p
Factor        1     16.38     16.38    15.36    0.004
Error         8      8.53      1.07
Total         9     24.92
                                  Individual 95% CIs For Mean
                                  Based on Pooled StDev
Level         N      Mean     StDev   ---------+---------+---------+-------
Premier       5    23.740     1.218   (------*------)
Super         5    26.300     0.806                    (------*------)
                                      ---------+---------+---------+-------
Pooled StDev =      1.033                    24.0      25.5      27.0
```

Data output 7.19

The first part of the output gives an ANOVA table in a fairly standard form (see the 'ideal' version in the example section). The highly significant *P* value of 0.004 means we reject the null hypothesis that the two cultivars have the same mean. However, it does not help us decide which group has the higher mean. The second section provides information about the two groups: number of observations (N), mean and standard deviation and then a rather primitive graphical representation of the two group means and 95% confidence intervals of the means. In this example the confidence intervals of the two groups do not overlap (confirming that they are significantly different from each other) and clearly '*Super*' has the greater mean.

Method 2. Input all the observations into a single column. Use a separate column for coded labels for the two groups. In this example you would call the first column 'Grain sz' and the second 'Cultivar'. You must use integers as codes for groups.

From the 'Stat' menu select 'ANOVA' then 'Oneway...'. Move the observed data into the 'Response:' box and the group codes into the 'Factor:' box. Don't worry about the comparisons button for only two groups, it will

Chapter 7
The tests 1: tests
to look at
differences

become useful when you have more than two groups. Click 'OK'. (Or type 'onew c1 c2' at the MTB> prompt in the session window.)

The output is identical to that from method 1 except that the two groups will be labelled with numbers rather than names.

Excel

Excel. There is no direct method unless you have installed the 'Analysis ToolPak add-in'.

1 Put the data into an Excel spreadsheet. This can be done in several ways: either exactly as in the example with one column for each cultivar and the names in the first row, or with all the data in a single column and an extra column having the names of the cultivars. (I will assume here that the data are in two columns A and B with labels in cells A1 and B1.)

2 Select the 'Tools' menu and select 'Data analysis...' (if this option does not appear you need to run Excel set-up and the 'Analysis ToolPak add-in'). Select 'Anova: Single Factor'.

3 A dialogue box will appear. If the cursor is flashing in the 'Input range' box you can select the cells you wish to use for the analysis by clicking and dragging in the main sheet. If you select from cell A1 to the end of the data '$A$1:$B$6' should appear in the box. Alternatively you can just type in 'A1:B6'. The option 'columns' should be selected because the data are indeed in two columns representing different groups. The tick-box 'Labels in the first

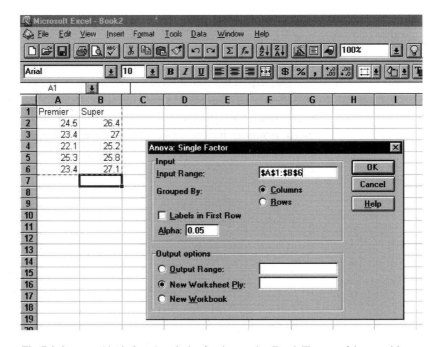

Fig. 7.4 One-way (single factor) analysis of variance using Excel. The area of the spreadsheet has been selected and is highlighted with a dashed line. Currently the output options will put the output onto a fresh sheet in the same workbook.

Chapter 7
The tests 1: tests
to look at
differences

row' should be checked as the cultivar names are in the first row. Leave the 'alpha' at 0.05 as this is the significance level which is chosen and $P<0.05$ is the usual level (Fig. 7.4).

4　At the bottom of the dialogue box is a section allowing you to determine where the output will appear. The default option is 'New Worksheet Ply' which means that the output will appear in a different sheet. If you want the output to appear on the same sheet as the data then you need to put a cell number in the 'Output range' box that will determine where the top left cell of the output will start.

5　Click on 'OK' and the following output will appear:

Anova: Single Factor

SUMMARY

Groups	Count	Sum	Average	Variance
Premier	5	118.7	23.74	1.483
Super	5	131.5	26.3	0.65

ANOVA

Source of Variation	SS	df	MS	F	P-value	F crit
Between Groups	16.384	1	16.384	15.3624	0.004422	5.317645
Within Groups	8.532	8	1.0665			
Total	24.916	9				

Data output 7.20

The first section gives summary information about the two groups of observations confirming the number of observations, sum, mean and variance.

The second section presents a rather standard ANOVA table: 'SS' is the sum of squares; 'df' degrees of freedom (there were 10 observations in all giving 9 total degrees of freedom and there were two groups giving one degree of freedom for the 'between groups' variation); 'MS' is the mean square (=SS/df) and F is the ratio of the variation between groups/within groups; 'P-value' is the important value as it shows the probability that the two cultivars have the same mean grain size. In this case $P=0.0044$ so we reject the null hypothesis that the two cultivars have the same mean grain size. The final value 'F-crit' is not usually quoted on an ANOVA table. It is the value of F required to achieve a $P=0.05$ with the degrees of freedom in this particular test. If alpha had been set at 0.01 in the ANOVA dialogue box then the 'F-crit' value would be 11.25863.

Mann–Whitney U

This test, also widely known as the Wilcoxon–Mann–Whitney test and less widely as the Wilcoxon rank sum W test, is the nonparametric equivalent of the independent samples t-test. It can only be used to test two groups.

Chapter 7
The tests 1: tests
to look at
differences

However, unlike the *t*-test and one-way ANOVA it does not make assumptions about homogeneity of variances or normal distributions. It is a typical 'rank' test, meaning that the raw data is converted into ranks before the test is carried out. The advantage of this is that it is ideal for situations where the highest value went off the scale or if extreme values are making the *t*-test undesirable.

The Mann–Whitney *U* test is slightly less powerful than a *t*-test or one-way ANOVA, but you are less likely to find a significant result when there is no real difference. However, the lack of assumptions it makes about the distribution of the data make it the preferred test in most cases.

An example

In the package illustrations for this test I use the same data set as for the *t*-test and one-way ANOVA (page 80).

SPSS

SPSS. As with the *t*-test all the data are placed into a single column and another column is used for the labelling of the groups using integer code numbers. Under the 'statistics' menu, choose 'nonparametric tests' and then select '2 Independent Samples...'. This will bring up a dialogue box for four different nonparametric tests. By default the 'Mann–Whitney *U* test' box should be selected. Put 'grain_sz' into the 'Test Variable List' box by selecting it and then moving it across. Then select 'cultivar' as the 'Grouping Variable'. It will appear as 'cultivar(? ?)'. You need to click the 'Define Range...' button and then input '1' in 'Group 1' and '2' in 'Group 2' before clicking 'Continue'. This will return you to the first dialogue box. You will see that 'cultivar(? ?)' is now 'cultivar(1 2)'. Click 'OK' to run the test.

The output should appear as follows:

```
- - - - - Mann-Whitney U - Wilcoxon Rank Sum W Test

    GRAIN_SZ
 by CULTIVAR


  Mean Rank    Cases

     3.20        5   CULTIVAR = 1
     7.80        5   CULTIVAR = 2
                --
                10   Total

                           Exact          Corrected for ties
      U          W      2-Tailed P       Z       2-Tailed P
     1.0       16.0       .0159      -2.4096       .0160
```

Data output 7.21

This confirms the test carried out, giving two different names for the same test! It also confirms that the variable 'grain_sz' was divided into two groups by the variable 'cultivar'. Next comes some summary information. The two

Chapter 7
The tests 1: tests
to look at
differences

groups of cultivar labelled 1 and 2 each have five cases (observations) giving 10 in total. The mean rank position of the two groups, with the smallest value given rank 1 and the largest rank 10, is also given. Even at this point it is clear that there is a difference between the two cultivars as the mean ranks are different.

Finally comes the output from the test itself. The first two columns, labelled U and W are the output values from the Mann–Whitney and Wilcoxon versions of the test respectively. Then the column labelled 'Exact 2-tailed P' gives the crucial *P* value. It is this value that you are interested in. Is it less than 0.05? In this example it is and we reject the null hypothesis that the two groups have the same median.

The last two columns give the output for yet another version of the test. This is a slightly more conservative version in the way it treats the ranking of tied values (when two data points have exactly the same value). In this particular case the *P* value it produces is very similar to the normal test. If there had been several ties then the difference might be greater. If you have a situation where the 'Exact 2-tailed P' is less than 0.05 and the 'Corrected for ties' version is greater then you should perhaps collect a little more data or repeat the experiment in exactly the same way as if you had a *P* value between 0.05 and 0.1.

MINITAB

MINITAB. Unlike the *t*-test there is only one way of inputting the data. Put the data into two columns, one for each group and label accordingly. From the 'Stat' menu select 'Nonparametrics' then 'Mann Whitney…'. In the dialogue box move the two columns into the 'First sample:' and 'Second sample:' boxes (it doesn't matter which way round they are). Leave the other settings at their defaults or '95.0' and 'not equal'. Click 'OK'. (Or type 'Mann-Whitney c1 c2' at the MTB> prompt in the session window.)

You get the following output:

```
Mann-Whitney Confidence Interval and Test

Premier    N =  5      Median =     23.400
Super      N =  5      Median =     26.400
Point estimate for ETA1-ETA2 is     -2.500
96.3 Percent C.I. for ETA1-ETA2 is (-4.301,-0.701)
W = 16.0
Test of ETA1 = ETA2  vs.  ETA1 ~= ETA2 is significant at 0.0216
The test is significant at 0.0212 (adjusted for ties)
```

Data output 7.22

The output confirms the test carried out. Then it gives some summary information about the two groups: the number of observations ('N') and the median value. The next two lines give information about the difference between the two groups and quote a range for the difference. Then comes a value for the test statistic 'W', given as 16.0 in the example. The final two lines are the most important. They give the probability of the two medians

Chapter 7
The tests 1: tests
to look at
differences

being equal. The first probability ('p=0.0216' in the example) does not account for tied data (two or more observations with exactly the same value) while the bottom line does. If there are several ties in the data then the second value will be higher and the second P value for the test becomes more conservative.

In the example both versions give a $P<0.05$ so we reject the null hypothesis that the two groups have the same median value (i.e. the grain size for the two cultivars is different). If the two significance results are either side of 0.05 then you will have to consider that a marginal result for the test suggests that you need to collect more data or measure with more precision to avoid tied data.

Excel

Excel. There is no direct method even with the 'Analysis ToolPak' installed.

Do the observations from more than two groups differ?

The groups can be either repeated (related) samples or they can be in-dependent. Repeated measures are considered first.

Repeated measures (a.k.a. related samples, matched samples)

This is an extension from paired data and occurs when a single individual or site is tested three or more times. A common example is in a 'before, during and after' design. Another possible use occurs when an individual is, or individuals of a clone are, divided and then subjected to three or more treatments. Two tests are considered below: the Friedman test; two-way/repeated measures ANOVA (page 98).

Friedman test (for repeated measures)

The Friedman test is a nonparametric analogue of a two-way ANOVA. It makes no assumptions about the distribution of the data only that it is measured on an ordinal scale. The Friedman is appropriate only if there is a single observation taken for each combination of factor levels. For repeated measures one of the factors must represent the repeat level, perhaps minutes, days or a measure of before, during and after a procedure. Then the second grouping variable will be a standard factor such as region, species or treatment type. The null hypothesis is that observations in the same group (factor level) have the same median values. If the null hypothesis is rejected it shows that at least two groups have different medians, although it does not show which groups they are. Inspection will reveal which are the likely candidate groups (i.e. those with the highest and lowest medians).

It is the case for both SPSS and MINITAB that the test has to be carried out twice if you wish to test both the conventional factor and whether the different sampling events have different median values.

[95]

Chapter 7
The tests 1: tests
to look at
differences

The Friedman test is much less powerful than an equivalent ANOVA test if the data is normally distributed but makes fewer assumptions about the data so it is 'safer'.

An example

The Friedman test could be used if the data comprises the number of cyano-bacterial cells in $1 \, mm^3$ of water from a group of six ponds, with samples taken on four different days and only one sample taken each time from each pond.

	Ponds					
Day	A	B	C	D	E	F
1	130	125	350	375	225	235
2	115	120	375	200	250	200
3	145	170	235	275	225	155
4	200	230	140	325	275	215

Note: there is only one observation for each pond/day combination.

SPSS

SPSS. Input the data using one column for each factor level (pond) in this example. Label the columns appropriately. Make sure that each row corresponds to a level (group) of the repeated measure (e.g. if different times make sure that all the 'before' measures are in the first row, 'during' in the second and so on).

From the 'Statistics' menu select 'Nonparametric tests' then 'K-related samples...'. In the dialogue box move all the columns containing data into the 'Test variables:' box. Make sure that the 'Friedman' box is checked. If you want summary statistics about each of the groups you should click the 'Statistics...' button and check 'Descriptives' and click 'continue' before you click 'OK'.

You should get the following output:

```
- - - - - Friedman Two-Way Anova

    Mean Rank     Variable

        1.50      pond_a
        2.50      pond_b
        4.25      pond_c
        5.38      pond_d
        4.25      pond_e
        3.13      pond_f

      Cases          Chi-Square        D.F.   Significance
        4             11.1786            5      4.80E-02
```

Data output 7.23

Chapter 7
The tests 1: tests
to look at
differences

The output confirms the test. Then gives the rather useless information about the mean rank of the observations in the different samples. On the last two lines there is the test output. First the number of repeated or related samples is confirmed. Then a test statistic is given followed by the degrees of freedom (one less than the number of groups being tested). Finally the P value is given, labelled as 'Significance'. If this value is less than 0.05 you reject the null hypothesis that the groups have the same median. In the example the P value is given in awkward scientific notation as '4.80E-02' which translates as 4.8×10^{-2} or 0.048. This value is just less than the critical 0.05 level. So we reject the null hypothesis that the ponds have the same median concentration of cyanobacteria.

To test another possible null hypothesis that the days all have the same median cell concentration would make the test a two-way analysis (see page 114).

MINITAB

MINITAB. Put the data into a single column. Then use the next two columns to put in integer codes for the two grouping variables. These codes must be integers but need not be consecutive.

From the 'Stat' menu select 'Nonparametric' then 'Friedman…'. In the dialogue box put the data in the 'Response' block, the conventional factor into the 'Treatment:' box and the repeated measure (perhaps a time as in the example) into the 'Block:' box. Click 'OK'. (Or, assuming the observations are in column 1, the main factor in column 2 and the repeat factor in column 3, type 'Friedman c1 c2 c3' at the MTB> prompt in the session window.)

You get the following output:

```
Friedman Test

Friedman test of Cells by Ponds blocked by Day

S = 11.18   d.f. = 5   p = 0.049
S = 11.26   d.f. = 5   p = 0.047 (adjusted for ties)

                   Est.     Sum of
    Ponds    N    Median     RANKS
      6      4    150.52      6.0
      7      4    166.35     10.0
      8      4    295.10     17.0
      9      4    293.85     21.5
     10      4    240.10     17.0
     11      4    192.19     12.5

Grand median =  223.02
```

Data output 7.24

Note: I coded the six ponds in the example with the numbers 6 to 11.

The output first confirms the test carried out. The first two rows give the results of the test, first with no correction for tied observations. The sample statistic for the test is 'S' and the 'd.f.' is the degrees of freedom (one less than the number of factor levels) there were six ponds in the data set so five degrees

Chapter 7
The tests 1: tests
to look at
differences

of freedom. Then there is a P value associated with the 'S' result. In the example the P value is just less than 0.05 so we reject the null hypothesis that the six ponds have the same median number of cells.

After the test results come some summary statistics about the groups giving number of observations ('N'), median and the total rank position of the observations in the whole data set (sum or RANKS). The 'Grand median' is the median value from the whole data set. By inspection you can determine that at least ponds 6 (A) and 9 (D) have significantly different medians as they are the extreme groups.

If you wish to determine whether the different sampling events have an effect on the median observations then you should repeat the test reversing the factors in the 'Treatment:' and 'Block' boxes. This makes the test into a 'two-way' analysis (see page 111). (Or type 'Friedman c1 c3 c2' at the MTB> prompt in the session window.)

Excel

Excel. There is no direct method even with the 'Analysis ToolPak' installed.

Repeated measures ANOVA

A two way analysis of variance may also be applied to this sampling design of only one measure for each combination of factor levels. ANOVA is a parametric test and therefore makes assumptions about the data. The data should be: continuous, normally distributed and with equal variances for each factor. In practice, it is often impossible to test whether this is true because there are so few observations. Therefore it is usual to make a value judgement about whether the data is likely to conform to the assumptions.

In the example used above in the Friedman test the data of cells/mm^3 is measured to the nearest 5 and is therefore discontinuous. However, there are more than 30 possible values so it would usually be accepted for ANOVA.

This test is only a special case of a two-way ANOVA where one of the factors defines the level of repeated sampling and the other the individuals or sampling sites. Treat as a two-way ANOVA (usually without replication), see page 110.

Independent samples

This is the more usual type of analysis and occurs when a single individual or site is measured or tested only once. There will, therefore be three or more totally separate groups of observations. Groups are often obvious: e.g. pigs, sheep, cows and horses. However, the distinction between the groups may be rather arbitrary: e.g. dividing samples by altitude bands or ranges of linear distance from a release point. Two tests are considered below: one-way analysis of variance and the Kruskal–Wallis test (page 106). If you get a significant result from one of these tests that is not the end of the story as it will not tell you which groups are different from which. A *post-hoc* test of some kind

Chapter 7
The tests 1: tests
to look at
differences

(meaning 'after this') is required to allow you to interpret the results. Some of the many *post-hoc* tests available are also described (page 108).

One-way ANOVA

The *t*-test is restricted to use with only two samples. When there are more than two groups the only sensible option for a parametric test is to use analysis of variance (ANOVA). This is still a very simple use of ANOVA but it works extremely well and is a test that every biologist should feel comfortable with.

ANOVA for three or more groups makes the same assumptions as for two groups: that the data is continuous; at least approximately normally distributed and that the variances of each group/sample are homogeneous. These assumptions should be tested before the test is carried out. The null hypothesis being tested is that each set of data has the same mean. (The way that ANOVA approaches this is to have a null hypothesis that the variation within groups is the same as variation between groups.)

An example

A researcher has grain size data from three cultivars and wishes to determine whether there are any differences between them. The null hypothesis is that all three will have the same mean. If there is a significant result (i.e. if $P<0.05$) this indicates that at least one pair of cultivars have different means, it does not tell you which one. For convenience, in this example I will use the same example data set as in the *t*-test but with an additional third sample of five observations from a cultivar called 'Dupa' which is coded as number 3.

Cultivar name	Cultivar code	Grain size (mg)
Premier	1	24.5
	1	23.4
	1	22.1
	1	25.3
	1	23.4
Super	2	26.4
	2	27.0
	2	25.2
	2	25.8
	2	27.1
Dupa	3	25.5
	3	25.7
	3	26.8
	3	27.3
	3	26.0

Chapter 7
The tests 1: tests
to look at
differences

Each of the three samples of five observations comes from a different cultivar of a crop plant.

If the data is analysed using ANOVA on a statistics package the output will not be directly suitable for presentation in a report. The ANOVA table that you would include in a write-up or publication should appear as something along the lines of the table below:

Source	d.f.	SS	MS	F	P
Cultivar	2	21.5	10.8	11.9	0.0014
Residual (error)	12	10.9	.91		
Total	14	32.4			

The best way to test that you are reading the output from your package correctly is to compare this table with the output obtained using the package.

SPSS

SPSS. As with many statistical tests all the data is placed into a single column and another column is used for the labelling of the groups. This may appear to double the effort required but it is a much easier system when it comes to multiway analysis. *Post hoc* tests are important when trying to interpret the output from ANOVA. They are considered later (page 108).

As with the two-sample tests there are at least three routes to this test in SPSS that give the same results but with rather different outputs. I will describe three below, considering two in detail.

Method 1. Under the 'statistics' menu choose 'compare means' and then 'One-way ANOVA...'. In the dialogue box that appears move 'grain_sz' into the 'Dependent list:' box by first highlighting it and then clicking on the appropriate move button. Next move 'cultivar' into the 'Factor' box. It will appear as 'cultivar(? ?)'. You need to click the 'Define Range...' button and then, assuming your three groups have been labelled 1, 2 & 3, input '1' in 'Minimum' and '3' in 'Maximum' before clicking 'Continue'. This will return you to the first dialogue box. You will see that 'cultivar(? ?)' now appears as 'cultivar(1 3)'. (Ignore the 'Post Hoc' button for the moment, although you should always select a *post hoc* test when there are more than two groups. I will consider *post hoc* testing below.) Click 'OK' to run the test.

The following will appear in the output window:

Chapter 7
The tests 1: tests
to look at
differences

```
               - - - - -  O N E W A Y  - - - - -

    Variable  GRAIN_SZ
    By Variable  CULTIVAR
                            Analysis of Variance

                                Sum of        Mean          F      F
        Source          D.F.    Squares       Squares      Ratio  Prob.

Between Groups            2      21.5093       10.7547     11.8792  .0014
Within Groups           12      10.8640         .9053
Total                   14      32.3733
```

Data output 7.25

This tells you that the observed data for 'grain_sz' has been grouped using 'cultivar' and then goes on to a standard ANOVA table. As is always the case in one-way ANOVA, the top line of the table is the important one; this is the variation between groups (i.e. between the three cultivars in this example). The second line is the variation within the groups that is being used as a comparison.

The first column is 'D.F.' or degrees of freedom. As there are three groups ('cultivars') there are two degrees of freedom between groups (i.e. $3-1=2$). There were five samples (with)in each group giving three lots of four degrees of freedom, and therefore 12 in total. Next come calculation columns: 'sum of squares' (often called SS) and 'mean squares' (MS). The mean square value is the sum of square value divided by the degrees of freedom. Finally comes the important bit; the F ratio. This is the mean square for between groups divided by that for within groups. If there is the same amount of variation between and within groups this will give an F ratio of exactly one. In this case the F ratio is 11.8792. This value could be looked up on an F table although SPSS gives you the P value anyway, labelling it as F prob. The value is 0.0014 indicating that *at least* two of the groups have means that are highly significantly different. There are three possible pairs with three groups: 1 & 2; 1 & 3; 2 & 3. It is the *post hoc* test that tells you which pairs are different from which *not the ANOVA itself*!

'Options...': one of the assumptions of ANOVA is that variances are equal. In the *t*-test in SPSS this is tested automatically. In ANOVA it is available under the 'Options...' button. Just check the box labelled 'Homogeneity of Variance' and click 'OK' before running the test. If you do this a little extra output appears after the ANOVA table.

```
Levene Test for Homogeneity of Variances

   Statistic    df1    df2    2-tail Sig.
     .6790        2     12        .526
```

Data output 7.26

The important value is given the label '2-tail Sig.'. The critical value is usually 0.05 and if the value given here is less than that ANOVA should not be used and a Kruskal–Wallis test should be considered instead (page 106).

Chapter 7
The tests 1: tests
to look at
differences

Method 2. Under the 'statistics' menu choose 'ANOVA Models' and then 'Simple Factorial…'. In the dialogue box that appears move 'grain_sz' into the 'Dependent:' box by first highlighting it and then clicking on the appropriate move button. Next move 'cultivar' into the 'Factor(s)' box. It will appear as 'cultivar(? ?)'. You need to click the 'Define Range…' button and then input '1' in 'Minimum' and '3' in 'Maximum' before clicking 'Continue'. This will return you to the first dialogue box showing 'cultivar(? ?)' as 'cultivar(1 3)'. Click 'OK' to run the test.

(Note that unlike Method 1 there is no option to run a Levene test for homogeneity of variances at the same time as running the ANOVA. For this reason you may prefer to use method 1. However, more complicated ANOVA tests are not possible under the 'Compare Means' menu so it may be better to become familiar with this method.)

```
* * *   A N A L Y S I S   O F   V A R I A N C E   * * *

        GRAIN_SZ
   by   CULTIVAR

        UNIQUE sums of squares
        All effects entered simultaneously

                            Sum of              Mean              Sig
Source of Variation         Squares    DF       Square      F     of F

Main Effects                21.509      2       10.755    11.879  .001
    CULTIVAR                21.509      2       10.755    11.879  .001

Explained                   21.509      2       10.755    11.879  .001

Residual                    10.864     12         .905

Total                       32.373     14        2.312


15 cases were processed.
0 cases (.0 pct) were missing.
```

Data output 7.27

As in method 1 this confirms that the data for 'grain_sz' have been grouped using 'cultivar' and then goes on to a rather less standard ANOVA table. The 'UNIQUE sums of squares' and 'All effects entered simultaneously' confirm standard calculation options that you will not want to change until you move on to very advanced statistics. This output is designed to cope with many factors and therefore has extra lines that appear totally superfluous for this rather simple situation. The really important line is repeated three times, although you should read along the line labelled 'cultivar'. This is the variation that was labelled 'between groups' in the method 1. The line labelled 'residual' corresponds to the one labelled as 'within groups'.

The first column is 'sum of squares', then 'DF' or degrees of freedom, then 'mean square' (the mean square value is the sum of square value divided by the degrees of freedom). As there are three cultivars they have two degrees of freedom. There were five samples within each cultivar giving three times four degrees of freedom and therefore 12 in total for 'residual'.

Chapter 7
The tests 1: tests
to look at
differences

Finally comes the important bit; the *F* ratio, labelled 'F' here. This is the mean square for 'cultivar' divided by that for 'residual'. The value of 'F' is 11.879. SPSS gives you the *P* value associated with this value of 'F' and these degrees of freedom and labels it 'Sig of F'. In biology we usually look for a value less than 0.05. Here the probability is 0.001 and indicates that at least two cultivars are highly significantly different from each other. It is the *post hoc* test that tells you which pairs are different from which *not the ANOVA itself*!

The output of this method should be revised along the lines of the table I gave at the top of this section before it is included in a write-up or report.

Method 3. A third method in SPSS is very similar to method 2, except that you select 'General Factorial…' instead of 'Simple Factorial…'. Follow the same instructions as for method 2 and you get output as follows:

```
* * * * * * A n a l y s i s   o f   V a r i a n c e -- design   1 * * * * * *

Tests of Significance for GRAIN_SZ using UNIQUE sums of squares
Source of Variation          SS        DF        MS          F   Sig of F

WITHIN+RESIDUAL            10.86        12       .91
CULTIVAR                  21.51         2     10.75        11.88      .001

(Model)                   21.51         2     10.75        11.88      .001
(Total)                   32.37        14      2.31

R-Squared =         .664
Adjusted R-Squared =  .608

- - - - - - - - - - - - - - - - - - - - - - - - - - - - - - - - - - - -
```

Data output 7.28

As with method 2 you are most interested in the line labelled 'cultivar' and the *P* value, labelled as 'Sig of F' at the end of the line. The rest of the table is similar to that from method 2. The last two lines, labelled with 'R-Squared' are part of a regression analysis and should be ignored here.

MINITAB

MINITAB. There are two routes to this test in MINITAB. You can either input all the data into a single column with a second column to code the groups or you can put each group into a separate column.

Method 1. Input all the data into a single column and use a second column to label the cultivars with integers (as in the example). Label the columns appropriately. From the 'Stat' menu select 'ANOVA' then 'Oneway…'. Move the observed data into the 'Response:' box and the group codes into the 'Factor:' box. The comparisons button gives you access to the *post hoc* tests that are considered separately. Click 'OK' to run the test. (Or type 'Oneway c1 c2' (assuming the data is in c1 and the group codes in c2) at the MTB> prompt in the session window.)

You get the following output:

```
One-Way Analysis of Variance

Analysis of Variance on Grain sz
Source      DF       SS        MS       F        p
Cultivar     2    21.509    10.755   11.88    0.001
Error       12    10.864     0.905
Total       14    32.373
                                    Individual 95% CIs For Mean
                                    Based on Pooled StDev
Level      N      Mean     StDev    ---------+---------+---------+------
  1        5     23.740    1.218    (-------*-------)
  2        5     26.300    0.806                      (-------*-------)
  3        5     26.260    0.764                      (-------*-------)
                                    ---------+---------+---------+------
Pooled StDev =    0.951                    24.0      25.2      26.4
```

Data output 7.29

The first part of the output confirms the test carried out and gives an ANOVA table in standard form (see the 'ideal' version in the example section). The highly significant P value (much less than 0.05) means we reject the null hypothesis that the cultivars have the same mean. However it does not help us decide which groups are different from which. The second section provides information about the three groups: number of observations ('N'), mean and standard deviation and then a rather primitive graphical representation of the three group means and 95% confidence intervals of the means. In this example the confidence intervals of the group 1 does not overlap with groups 2 and 3 (confirming that they are significantly different from each other). The *post hoc* test will confirm this inspection of the data (page 108).

Method 2. There is no option for *post hoc* testing with this method making method 1 far superior. Input the data with one column for each of the groups (cultivars in the example) and label the columns appropriately. From the 'Stat' menu select 'ANOVA' then 'Oneway (unstacked)...'. In the dialogue box move all groups into the 'Responses (in separate columns):' box. Then click 'OK'. The output is identical to method 1 but with the column names appearing instead of the factor levels.

Excel

Excel. There is no direct method unless you have installed the 'Analysis ToolPak add-in'. The method is essentially the same as for two groups.

1   Input the data onto an Excel spreadsheet. This can be done in several ways: either with one column for each cultivar and the names in the first row, or as in the example with the data from one cultivar in one row with an extra column having the names of the cultivars. (I will assume here that the data are in three columns A, B and C with names of the cultivars in cells A1, B1 and C1.)

2   Select the 'Tools' menu and select 'Data analysis...' (if this option does

Chapter 7
The tests 1: tests
to look at
differences

not appear you need to run Excel set-up and add the 'Analysis ToolPak add-in'). Select 'Anova: Single Factor'.

3 A dialogue box will appear. If the cursor is flashing in the 'Input range' box you can select the cells you wish to use for the analysis by clicking and dragging in the main sheet. If you select from cell A1 to the end of the data '$A$1:$C$6' should appear in the box. Alternatively you can just type in 'A1:C6'. The option 'columns' should be selected because the data are indeed in two columns representing different groups. The tick-box 'Labels in the first row' should be checked as the cultivar names are in the the first row. Leave the 'alpha' at 0.05 as this is the significance level which is chosen and $P < 0.05$ is the usual level used in biology.

4 At the bottom of the dialogue box is a section allowing you to determine where the output will appear. The default option is 'New Worksheet Ply' which means that the output will appear in a different sheet. If you want the output to appear on the same sheet as the data then you need to put a cell number in the 'Output range' box that will determine where the top left cell of the output will start.

5 Click on 'OK' and the following output will appear:

Anova: Single Factor

SUMMARY

Groups	Count	Sum	Average	Variance
Premier	5	118.7	23.74	1.483
Super	5	131.5	26.3	0.65
Dupa	5	131.3	26.26	0.583

ANOVA

Source of Variation	SS	df	MS	F	P-value	F crit
Between Groups	21.50933	2	10.75467	11.87923	0.001428	3.88529
Within Groups	10.864	12	0.905333			
Total	32.37333	14				

Data output 7.30

The important value is the 'P-value'. If this is less than 0.05 we reject the null hypothesis that all three cultivars have grains with the same mean size. In this case the value is well below that level at $P = 0.0014$ so we can say that the result is highly significant. However, this just gives the probability that two of the cultivars have different means — it does not show which ones. Inspection of the 'Summary' table shows that '*Super*' and '*Dupa*' have very similar means but '*Premier*' is different from them. There is no *post-hoc* test available in Excel.

See the section on two groups above for a description of the rest of the output (page 92).

Chapter 7
The tests 1: tests
to look at
differences

Kruskal–Wallis test

This test is the nonparametric equivalent of the one-way analysis of variance and has a null hypothesis that all samples are taken from populations with the same median. It can be used to test any number of groups. However, unlike one-way ANOVA it does not make assumptions about homogeneity of variances or normal distributions. It is a typical 'rank' test, meaning that the raw data are converted into ranks before the test is carried out. The advantage of this is that it is ideal for situations where the highest value went off the scale or if extreme values are present as these have a disproportionate influence on the results of parametric tests. This test *may* be used when there are only two samples, but the Mann–Whitney *U* test (page 92) is more powerful for two samples and should be preferred.

This test is somewhat less powerful than one-way ANOVA, but you are less likely to find a significant result when there is no real difference (i.e. the probability of a Type I error is decreased).

Example

I will be using the same data set as for the one-way ANOVA (page 99) so the results from the two tests may be compared.

SPSS

SPSS. Under the 'Statistics' menu, choose 'Nonparametric tests' and then select 'k Independent Samples…'. This will bring up a dialogue box for two nonparametric tests. By default the Kruskal–Wallis test box should be selected. Put 'grain_sz' into the 'Test Variable List' box by selecting it and then moving it across. Then select 'cultivar' as the 'Grouping Variable'. It will appear as 'cultivar(? ?)'. You need to click the 'Define Range…' button and then input '1' in 'Minimum' and '3' in 'Maximum' before clicking 'Continue'. This will return you to the first dialogue box. You will see that 'cultivar(? ?)' is now 'cultivar(1 3)'. Click 'OK' to run the test.

The output appears as follows:

```
- - - - - Kruskal-Wallis 1-Way Anova

     GRAIN_SZ
  by CULTIVAR


   Mean Rank    Cases

        3.20       5    CULTIVAR = 1
       10.40       5    CULTIVAR = 2
       10.40       5    CULTIVAR = 3
                  --
                  15    Total

                                            Corrected for ties
  Chi-Square    D.F.  Significance    Chi-Square    D.F.  Significance
    8.6400        2         .0133       8.6555         2         .0132
```

Data output 7.31

Chapter 7
The tests 1: tests
to look at
differences

This confirms the test carried out, and even calls it a 1-way ANOVA even though it isn't! Next comes confirmation that the variable 'grain_sz' was divided into groups by the variable 'cultivar'. Summary information about the three samples follows. The groups of 'cultivar' labelled '1', '2' and '3' each have five cases (observations) giving 15 in total. The mean rank position of the groups (the smallest value of the 15 is given rank one and the largest rank 15), is also given. Even at this point it is clear that there is a difference between the cultivars as the mean rank of group '1' is very different to groups '2' and '3'.

Finally comes the output from the test itself in two versions. The first three columns are the basic version of the test giving a 'Chi-square' value an appropriate degree of freedom (there are three groups, giving two degrees of freedom) and then a *P* value, labelled as 'Significance'.

Then comes a slightly more conservative version of the test in the way it treats the ranking of tied values (when two items of data have exactly the same value). In this case the *P* value it produces is very similar to the normal test. If there had been several ties then the difference might be greater. If you have a situation where the 'Significance' is less than 0.05 and the 'Corrected for ties' version is greater then you should perhaps collect a little more data or repeat the experiment in exactly the same way as if you had a *P* value between 0.05 and 0.1.

MINITAB

MINITAB. Input all the data into a single column and use a second column to label the cultivars with integers (as in the example). Label the columns appropriately. From the 'Stat' menu select 'Nonparametrics' then 'Kruskal–Wallis...'. In the dialogue box move the observed data into the 'Response:' box and the group codes into the 'Factor:' box. Click 'OK' to run the test. (Or type 'Krus c1 c2' (assuming the data is in c1 and the group codes in c2) at the MTB> prompt in the session window.)

You get the following output:

```
Kruskal-Wallis Test

LEVEL      NOBS     MEDIAN  AVE. RANK    Z VALUE
   1         5      23.40        3.2      -2.94
   2         5      26.40       10.4       1.47
   3         5      26.00       10.4       1.47
OVERALL     15                   8.0

H = 8.64   d.f. = 2  p = 0.014
H = 8.66   d.f. = 2  p = 0.013 (adjusted for ties)
```

Data output 7.32

The output confirms the test used. Then gives some summary information for each of the groups giving their integer codes. Number of observations is 'NOBS' here, the median and mean rank are also given. The 'z-value' is used in the test calculation.

Chapter 7
The tests 1: tests
to look at
differences

The last two lines give the test result. There are two versions depending on how they treat tied observations in the data. 'H' is a test statistic, 'd.f.' is the degrees of freedom (one less than the number of groups). Finally the *P*-value, labelled 'p', is given. In the example it is far less than the critical 0.05 level and we reject the null hypothesis that the three groups have the same median. There is no *post hoc* test available to determine which group is different from which although this can be done by inspection of the summary statistics.

Excel

Excel. There is no direct method.

Post hoc *testing: after one-way ANOVA*

One of the commonest errors of omission I see is a one-way ANOVA carried out on, say, four groups with a *P* value well under 0.05 and then no further investigation of which groups are different from which. It is important to realize that a significant result in the ANOVA will only show that at least one pair of the groups is significantly different. It does not identify which pair(s). When there are three groups that is only three possible pairs, with four groups that rises to six pairs and with five groups there are ten. *Post hoc* tests help you to make sense of this large number of possible comparisons by actually identifying which groups are significantly different from which.

The only problem is the number of methods at your disposal for this job is overwhelming. For instance in SPSS you are offered a choice of seven different methods for answering the same question and these are only a subset of the number available.

The two tests I suggest you look for, and they will almost invariably give you the same results, are the least significant difference (LSD) test (a.k.a. Fisher's LSD test) and the Student Newman Keuls (SNK) test.

It is conventional to show the results of a significant ANOVA by arranging the groups into ascending order of mean and then drawing lines under groups that are *not* significantly different. Therefore for the example I have been using here:

```
Cultivar 1    Cultivar 3    Cultivar 2
              _____
```

SPSS

SPSS. If you have followed method 1, 'Compare means' then 'one-way ANOVA', then you must select 'Post Hoc' in the dialogue box before you choose 'OK'. This will offer you a choice of seven *post hoc* tests, all currently unselected. Select either 'least significant difference' or 'Student Newman Keuls' then click on 'Continue' to return you to the main dialogue box.

The output from each test is similar. I have shown that from the LSD test as follows:

Chapter 7
The tests 1: tests
to look at
differences

```
- - - - - O N E W A Y - - - - -

        Variable   GRAIN_SZ
    By Variable    CULTIVAR

Multiple Range Tests:  LSD test with significance level .05

The difference between two means is significant if
   MEAN(J)-MEAN(I)   >= .6728 * RANGE * SQRT(1/N(I) + 1/N(J))
   with the following value(s) for RANGE: 3.08

   (*) Indicates significant differences which are shown in the lower triangle

                                 G G G
                                 r r r
                                 p p p

                                 1 3 2
       Mean        CULTIVAR

     23.7400       Grp 1
     26.2600       Grp 3          *
     26.3000       Grp 2          *

 Homogeneous Subsets (highest and lowest means are not significantly different)

Subset 1

Group        Grp 1

Mean        23.7400
- - - - - -

Subset 2

Group        Grp 3          Grp 2

Mean        26.2600        26.3000
- - - - - - - - - - - - - - -
```

Data output 7.33

This output confirms the test in the same way as the one-way ANOVA itself. Then it gives an idea of the way the calculation works before getting to the result you are interested in which is presented in a matrix where the groups are put into mean order and upper right half is always blank. An asterisk shows which pairs are significantly different (i.e. the *P* value is less than 0.05).

In this case the two asterisks show that group '1' is significantly different from '2' and '3' but they are not different from each other. This result is then confirmed with further output dividing the data into two subsets where the groups are not significantly different. 'Subset 1' contains only group '1' while 'subset 2' has groups '2' and '3'.

MINITAB

MINITAB. Before you click 'OK' to start a one-way test click the 'Comparisons…' button. This takes you to a dialogue box with four different tests. Select the 'Fisher's individual error rate:' option and make sure that the number in the box is a 5, this is the percentage significance level so 5 translates as critical *P* value of 0.05.

When you run the test you get the following output after the standard ANOVA table:

```
                                        Individual 95% CIs For Mean
                                        Based on Pooled StDev
    Level      N       Mean      StDev   ----------+---------+---------+------
      1        5      23.740     1.218   (-------*-------)
      2        5      26.300     0.806                     (-------*-------)
      3        5      26.260     0.764                     (-------*-------)
                                        ----------+---------+---------+------
  Pooled StDev =     0.951                    24.0      25.2      26.4

  Fisher's pairwise comparisons

     Family error rate = 0.116
  Individual error rate = 0.0500

  Critical value = 2.179

  Intervals for (column level mean) - (row level mean)

                 1          2

      2      -3.8713
             -1.2487

      3      -3.8313    -1.2713
             -1.2087     1.3513
```

Data output 7.34

The graphical display of the mean and confidence interval for the means by groups shows quite clearly in this case that factor level '1' is very different from '2' and '3'. The 'Fisher's pairwise comparisons' section first confirms the error rate chosen 0.05 in the 'Individual error rate' and then gives the critical mean difference required for two factor levels (groups) to be deemed different at the 0.05 significance level. In the example group means have to be 2.179 apart to be deemed significantly different.

Finally comes a small table of the group comparisons showing the 95% confidence range for the mean difference between the groups. If both values are negative then the column level is significantly *lower* than the row level. If both are positive then the column level is significantly *higher* than the row level. If one is positive and one negative then the column and row levels are not significantly different. In the example level 1 '*Premier*' is significantly lower than both level 2 ('*Super*') and level 3 ('*Dupa*') while level 2 is not significantly different from level 3. If i is the number of groups then there will be $i(i-1)/2$ comparisons (i.e. for 3 groups, 6 for 4 groups, 45 for 10 groups) so the comparisons in this form become increasingly difficult to interpret.

Excel

Excel. No direct methods are available.

There are two independent ways of classifying the data

If, for each observation, you have two factors (different ways of subdividing the data into groups) and the factors are independent of each other (i.e. there is no way the level of one factor can be deemed to be affected by the level of another) then there are several tests for analysing the null hypothesis that all factor levels have the same mean. For example, you intend to record the beak

Chapter 7
The tests 1: tests
to look at
differences

widths of house sparrows in several different towns: your observations will be the measurement of beak width, one factor might be the sex of the bird and the other the town where it was collected.

In addition there may be a further null hypothesis that there is no interaction between the two factors under investigation. It is important to realize that interaction may only be investigated when there is more than one observation for each factor combination.

One observation for each factor combination (no replication)

There will be many circumstances where you wish to test the effect of two factors but are only able to take a very small number of observations. If you have only one observation for each factor combination then there are still tests you can perform to test the null hypothesis that each factor level has the same mean or median. The main difference in the interpretation of the results of these tests, when compared to tests with replication is that there is no null hypothesis that there is no interaction between the two factors.

An example

A trial of six different blends of fertilizers (coded with the letters U to Z) has been carried out on linseed crops on four different farms (coded 1 to 4). Factor 1 is the fertilizer and factor 2 the farm. On each farm six fields were used in the trial so it was only possible to use each of the six fertilizer blends once on each farm. The crop yields of the linseed are given in the table.

Farm	Fertilizer blend					
	U	V	W	X	Y	Z
1	1130	1125	1350	1375	1225	1235
2	1115	1120	1375	1200	1250	1200
3	1145	1170	1235	1175	1225	1155
4	1200	1230	1140	1325	1275	1215

Friedman test

The Friedman test is a nonparametric analogue of a two-way ANOVA (see below) that can only be used when there is a single observation for each factor combination. The null hypothesis is that the median values of each factor level is the same. It does not make assumptions about the distribution of the data and can be used in any circumstance when the data is at least on an ordinal

Chapter 7
The tests 1: tests
to look at
differences

SPSS

scale. However, the test is rather conservative. Another problem is that in many packages the test has to be carried out twice: once to compare the rows and once for the columns.

SPSS. To test the null hypothesis that there is no difference between the columns (fertilizer blends in the example). Input the data into SPSS in exactly the same format as it is in the table of raw data. You can either label the columns with the code letters U–Z or leave them as the default 'var00001' etc.

From the '<u>S</u>tatistics' menu choose '<u>N</u>onparametric tests' and then 'K related samples…'. In the dialogue box that appears move all the columns into the 'test variables' box. Make sure that 'Friedman' is checked and click 'OK'.

This output should appear:

```
- - - - - Friedman Two-Way Anova


   Mean Rank    Variable

        1.50    U
        2.50    V
        4.50    W
        4.88    X
        4.50    Y
        3.13    Z

        Cases        Chi-Square       D.F.    Significance
          4            10.3214          5       6.66E-02
```

Data output 7.35

The first section of the output confirms the test used then shows the mean rank position of the yield data for each of the six fertilizer blends (these are 1 as the lowest on each farm and 6 as the highest). So the mean rank of blend 'U' is 1.5 on the four farms (it was lowest twice and second lowest twice).

The second section of the output shows that there were 4 farms ('Cases'), that the test output was 10.3214, there are 5 degrees of freedom 'D.F.' and that the P value ('Significance') is given in scientific notation as 6.66E-02 (i.e. 6.66×10^{-2} or 0.0666). This value is very close to the critical significance level of 0.05. So with such a conservative test we should seriously consider further trials.

You may wish to stop at this point or continue to step 2 depending on whether you are interested in the differences between farms or not.

To test the null hypothesis that there is no difference between farms. The data needs to be transposed so that each farm is in a different column. This is very easy in SPSS: select 'Data' then 'Transpose'. Move the column labels for each of the six fertilizer blends in the box labelled 'Variable(s)' and click 'OK'.

Carry out the Friedman test as in step 1 and you get following output:

Chapter 7
The tests 1: tests
to look at
differences

```
- - - - - Friedman Two-Way Anova


   Mean Rank    Variable

       2.75     VAR001
       2.17     VAR002
       1.92     VAR003
       3.17     VAR004

       Cases           Chi-Square           D.F.    Significance
         6               3.4500               3       3.27E-01
```

Data output 7.36

The details are as before. The most important number is the 'Significance' value of 3.27E-01 which translates as $P=0.327$. We can therefore accept the null hypothesis that there is no difference between yields on different farms. In the example the variable names were left as the default SPSS names. These can be replaced easily by double-clicking on the column heading and changing the variable attributes in the dialogue box that appears.

MINITAB

MINITAB. Input the observations into a single column. Use two further columns for the two grouping variables (factors) replacing the labels for factor levels with integers. Label the columns appropriately.

From the 'Stat' menu select 'Nonparametrics' then 'Friedman...'. Put the observed data column into the 'Response:' box. Then put one of the factor columns into the 'Treatment:' box and the other into the 'Block:' box. If you are only really interested in one of the two factors then that should be in the 'Treatment:' box. Click 'OK' to run the test. (Or, assuming the observations are in column 1, and the two factors in columns 2 and 3, type 'Friedman c1 c2 c3' at the MTB> prompt in the session window.)

You get the following output:

```
Friedman Test

Friedman test of Yield by Fertiliz blocked by Farm

S = 10.32   d.f. = 5   p = 0.068
S = 10.40   d.f. = 5   p = 0.066 (adjusted for ties)

                       Est.     Sum of
Fertiliz      N       Median     RANKS
     1        4       1147.1      6.0
     2        4       1161.2     10.0
     3        4       1296.3     18.0
     4        4       1255.8     19.5
     5        4       1241.3     18.0
     6        4       1193.3     12.5

Grand median   =    1215.8
```

Data output 7.37

The output confirms the test carried out then the name of the column containing the data with the factor being tested ('Fertiliz') in the example mentioned next. Then come two lines with the test statistic 'S', the degrees of

Chapter 7
The tests 1: tests
to look at
differences

freedom 'd.f.' (one less than the number of factor levels) and the *P* value associated with the test statistic. There are two versions of the test depending on how tied (equal) values are dealt with. The *P* values will be similar unless there a lot of tied observations in the observed data. Use the *P* value in the second row (adjusted for ties). In this case the *P* value is close to the critical 0.05 level so that although we accept the null hypothesis that there is no difference in median yield at different fertilizer levels we would consider further investigation. After the test comes some summary information about the different factor levels giving number of observations ('N'), the estimated median yield for each factor level and then the mean rank position of the observations in each factor level.

This completes investigation of one factor. To investigate the other factor: go to the 'Stat' menu, select 'Nonparametrics' then 'Friedman...' and swap the column names in the 'Block:' and 'Treatment:' boxes. (Or type 'Friedman c1 c3 c2' at the MTB> prompt.)

For the example data this gives:

```
Friedman test of Yield by Farm blocked by Fertiliz

S = 3.45   d.f. = 3   p = 0.328
S = 3.51   d.f. = 3   p = 0.320 (adjusted for ties)

                  Est.     Sum of
    Farm    N    Median    RANKS
     1      6    1232.3    16.5
     2      6    1213.6    13.0
     3      6    1193.0    11.5
     4      6    1255.5    19.0

Grand median  =   1223.6
```

Data output 7.38

Output is the same as for the first factor. In the example the *P* value is well above 0.05 so we accept the null hypothesis that farms have the same median yield.

There is no way to determine the significance of any interaction between the two factors.

Excel

Excel. There is no direct method.

Two-way ANOVA (without replication)

It may seem slightly odd that a statistical test that relies on the comparison of variation can be used when there is only one observation for each factor level combination but the test is perfectly valid. The two grouping variables may be set by the experimenter (e.g. different concentrations) or 'naturally' occurring (e.g. different sites). The assumptions of the test are: that the observed data are continuous; are approximately normally distributed; that the data would have about the same variance in each factor combination and that the grouping variables have at least two levels each that can be coded into integers.

Chapter 7
The tests 1: tests
to look at
differences

Obviously with only one observation in each factor combination there is no scope for testing the data against a normal distribution. Therefore it is up to the tester to use common sense or previous knowledge about the data under investigation.

Example

The same example data set as for the Friedman test is used in the illustrations below. The data are crop yields that are rounded (therefore discontinuous) but there are clearly far more than 30 possible values so the assumption of continuous data can be accepted. The normal distribution might be more difficult as observations such as crop yields are often skewed to the right. However, in this case we accept that the raw data are suitable for the ANOVA.

The results of the ANOVA should be presented in table form:

Factor	d.f.	SS	MS	F ratio	P value
Farm	3	11071	3690	0.797	0.515
Fertilizer blend	5	59763	11953	2.58	0.071
Error	15	69479	4631		

Where d.f. is degrees of freedom, SS is sum of squares, MS is mean square (MS = SS/d.f.), F ratio is the important statistic (F ratio = MS factor/MS error). Error is often referred to as residual.

Note that there are 23 degrees of freedom in total in this example. There were 24 observations in the total data set so that is correct. It is always sensible to check the number of degrees of freedom in the output to see if it matches your expectations.

There is no possibility of examining the interaction between the two factors as there is no replication. The interaction is an important part of the power of analysis of variance where there is replication.

In the example the P value associated with the fertilizer blends was close to the critical 0.05 level. In a report this could be highlighted in the text as follows: 'There was no significant difference in yield between farms ($F_{3,15}=0.797$, $P>0.1$) but there was an indication that yields varied between fertilizer blends ($F_{5,15}=2.58$, $P=0.071$)'. The subscripted numbers quoted with the F ratios are the degrees of freedom.

SPSS. Input the observed data into a single column. Use two separate columns to code the observations with integers to represent the factor levels. The integers do not have to be in sequence or start at 1. Label all the columns

[115]

appropriately. In the example there will be a 'Yield' column, a 'Farm' column (containing the code numbers 1–4) and 'Fertiliz' column with number codes replacing the letter codes from the original data.

There are now at least two methods for carrying out the test. They have rather different outputs so I will consider both. Method 2 allows for more additional output while method 1 is slightly more straightforward to implement.

Method 1. From the 'Statistics' menu select 'ANOVA models' then 'Simple factorial...'. In the dialogue box move the observations into the 'Dependent:' box and the two factors into the 'Factor(s):' box. For each factor in turn you must define the range of possible values for factor levels. Select the first factor, click the 'Define range...' button, then in the box put the value of the lowest factor level in the 'Minimum' box and the highest in the 'Maximum' box (in the example the factor 'Farm' would have a 1 in 'Minimum' and 4 in 'Maximum'). Click continue and return to the main dialogue box.

Before you carry out the test you must click on the 'Options...' button and bring up the options box. In the section marked 'Maximum interactions' select 'none' then click 'continue'. If you do not do this step the analysis will not provide any significance values! Once back to the original dialogue box click 'OK' to run the test.

You will get the following output:

```
* * *   A N A L Y S I S   O F   V A R I A N C E   * * *

              YIELD
         by   FARM
              FERTILIZ

              UNIQUE sums of squares
              All effects entered simultaneously


                              Sum of              Mean             Sig
Source of Variation           Squares    DF       Square      F    of F

Main Effects                  70833.333   8       8854.167   1.912  .133
     FARM                     11070.833   3       3690.278    .797  .515
     FERTILIZ                 59762.500   5      11952.500   2.580  .071

Explained                     70833.333   8       8854.167   1.912  .133

Residual                      69479.167  15       4631.944

Total                        140312.500  23       6100.543
```

Data output 7.39

The output first confirms the test and then gives the name of the column containing the observations and then the factor names. The next two lines relate to the fact that the ANOVA is using standard assumptions. Then comes the ANOVA table itself. There is more information that you actually need here (compare to the 'ideal' output table given in the description of the example). The important lines are for the factor variables, called 'Main Effects' in SPSS ('Farm' and 'Fertiliz' in the example). The 'DF' column gives the degrees of freedom which is one less than the number of factor levels (in the example

Chapter 7
The tests 1: tests
to look at
differences

there were four farms so the factor 'Farm' has 3 d.f.). The sum of squares and mean square are given, mean square is sum of squares/degrees of freedom. The last two columns give the important information. The 'F' column is the F ratio (factor mean square/residual mean square). The P value is labelled 'Sig of F' and given in the last column. In the example the P value for 'Farm' is well above the critical 0.05 level so we accept the null hypothesis that all farms have the same mean yield. However, the P value for 'Fertiliz' is 0.071, quite close to the critical level. We still accept the null hypothesis, but should note that the P value is close to the critical level in any report.

In this ouput it is best to totally ignore the lines labelled 'Main Effects' and 'Explained' as they are produced by combining the values for the two factors.

Method 2. From the 'Statistics' menu select 'ANOVA models' then 'General factorial…'. In the dialogue box move the observations into the 'Dependent:' box and the two factors into the 'Factor(s):' box. For each factor in turn you must define the range of possible values for factor levels. Select the first factor, click the 'Define range…' button, then in the box put the value of the lowest factor level in the 'Minimum' box and the highest in the 'Maximum' box (in the example the factor 'Fertiliz' would have a 1 in 'Minimum' and 6 in 'Maximum'). Click 'continue' and return to the main dialogue box.

Before you run the test you must move to the model screen to suppress interaction terms in the output. Click the 'Model…' button. In the model screen select the 'Custom' option. Then from the pull down menu between the two large boxes select 'Main effects'. Then highlight the two factors that should be in the box on the left and click the move button in the middle to copy them into the right hand 'Model:' box. This sets up a statistical test that just has the two factors as main effects and nothing else. Click 'continue'.

Back in the main dialogue box you can either click 'OK' straight away or move first into the 'Options…' box where you can choose to display summary statistics for each of the variables. If you do this it can make interpretation of the output easier.

You get several pages of output; here is the important page:

```
* * * * * * A n a l y s i s   o f   V a r i a n c e -- design   1 * * * * * *

Tests of Significance for YIELD using UNIQUE sums of squares
Source of Variation         SS        DF       MS        F   Sig of F

WITHIN+RESIDUAL         69479.17      15    4631.94
FARM                    11070.83       3    3690.28      .80     .515
FERTILIZ                59762.50       5   11952.50     2.58     .071

(Model)                 70833.33       8    8854.17     1.91     .133
(Total)                140312.50      23    6100.54

R-Squared =                .505
Adjusted R-Squared =       .241
```

Data output 7.40

The first line confirms the test and then the variable that is being tested. The table is slightly different to the 'ideal' as it has some extra lines. The

Chapter 7
The tests 1: tests
to look at
differences

important lines are those with the names of the two factors. The five columns of the table give the standard SS (sum of squares), d.f. (degree of freedom), MS (mean square), *F* ratio (labelled as 'F') and a *P* value (labelled 'Sig of F') associated with the *F* ratio and degrees of freedom. This is the value you compare with the critical level of 0.05. If *P* > 0.05 then you must accept the null hypothesis that different levels of the factor have the same mean. In this example the *P* value for 'Farm' is well above 0.05 but that for 'Fertiliz' is, at 0.071, close to the critical level. This suggests that perhaps further investigation is required.

The line labelled 'Within and residual' gives the error or residual line for a conventional ANOVA table. The line labelled '(Model)' combines the values for the two factors while the '(Total)' line sums both factors and residuals. Finally values are given for 'R-squared' and 'Adjusted R-squared' that give an idea of the proportion of the variation in the observations accounted for by the factors.

MINITAB

MINITAB. There are at least two ways of carrying out this test in MINITAB. I will only discuss one here as it gives the most useful output.

Input all the data into a single column and then use the next two columns for the factors replacing factor level names with integer codes. Label the columns appropriately. From the 'Stat' menu choose 'ANOVA' then 'General linear model...'. Move the data column into the 'Responses:' box and the two factor columns into the 'Model:' box. Click 'OK' to run the test. (Or type 'GLM c1 = c2 c3' at the MTB> prompt in the session window.)

If you used the example you get the following output:

```
General Linear Model

Factor    Levels Values
Fertiliz     6    1    2    3    4    5    6
Farm         4    1    2    3    4

Analysis of Variance for Yield

Source    DF    Seq SS    Adj SS    Adj MS      F      P
Fertiliz   5     59763     59763     11953   2.58   0.071
Farm       3     11071     11071      3690   0.80   0.515
Error     15     69479     69479      4632
Total     23    140312

Unusual Observations for Yield

Obs.    Yield      Fit  Stdev.Fit  Residual   St.Resid
 12    1140.00  1289.58    41.68    -149.58     -2.78R

R denotes an obs. with a large st. resid.
```

Data output 7.41

The output confirms that you have carried out a 'General linear model' a general term for a family of statistical tests that includes analysis of variance. Then it confirms the factors that have been used, how many levels there are for each factor and then the code numbers used for the factor levels (it pays to

Chapter 7
The tests 1: tests
to look at
differences

check these in case you made a mistake typing in the numbers!). Then comes the ANOVA table itself stating the name of the observation variable (Yield in the example). The table has the usual columns except that it adds an additional SS column (with exactly the same numbers in as the usual SS column). 'Source' refers to the source of variation, 'DF' is degrees of freedom, 'SS' is sum of squares, 'MS' is mean square, 'F' is the *F* ratio and 'P' the *P* value. Compare the output to the ideal table I used in the description of the example. In the example the *P* value for 'Farm' is well above 0.05 so we accept the null hypothesis but the *P* value for 'Fertiliz' is close to 0.05 so we may consider further investigation.

Finally comes a list of unusual observations. 'Obs.' tells you which row number is unusual. These can often be typing errors so you should check the list against the original data set.

Excel

Excel. There is no direct method unless you have installed the 'Analysis ToolPak add-in'.

1 Put the data into an Excel spreadsheet in exactly the format it is in the example with the appropriate labels or code letters as column and row labels.

2 Select the 'Tools' menu and select 'Data analysis...' (if this option does not appear you need to run Excel set-up and all the 'Analysis ToolPak add-in'). Select 'Anova: Two Factor Without Replication'.

3 A dialogue box will appear. If the cursor is flashing in the 'Input range' box you can select the cells you wish to use for the analysis by clicking and dragging in the main sheet. If you select from cell A1 to the end of the data '$A$1:$G$5' should appear in the box. Alternatively you can just type in 'A1:G5'. The tick-box 'Labels' should be checked as these are included. Leave the 'alpha' at 0.05 as this is the significance level which is chosen and $P<0.05$ is the usual level.

4 At the bottom of the dialogue box is a section allowing you to determine where the output will appear. The default option is 'New Worksheet Ply' which means that the output will appear in a different sheet. If you want the output to appear on the same sheet as the data then you need to put a cell number in the 'Output range' box that will determine where the top left cell of the output will start.

5 Click on 'OK' and the output overleaf will appear.

```
Anova: Two-Factor Without Replication

SUMMARY      Count      Sum      Average        Variance
     1         6        7440         1240           11180
     2         6        7260         1210            9230
     3         6        7105     1184.167        1384.167
     4         6        7385     1230.833        4054.167

U              4        4590        1147.5            1375
V              4        4645      1161.25         2606.25
W              4        5100         1275        11816.67
X              4        5075      1268.75        9322.917
Y              4        4975      1243.75        572.9167
Z              4        4805      1201.25         1156.25


ANOVA
Source of      SS        df         MS           F          P-value      F crit
Variation
Rows       11070.83       3      3690.278     0.796702      0.51467    3.287383
Columns     59762.5       5       11952.5     2.58045       0.070826   2.901295
Error      69479.17      15      4631.944

Total      140312.5      23
```

Data output 7.42

The first section summarizes the data for each factor level of the two factors in turn. 'Count' is the number of observations and 'Average' the mean.

In the second section of the output there is a conventional ANOVA table. The 'Source of Variation' refers to the two factors. 'Rows' are farms and 'Columns' are fertilizer blends in this example. 'SS' is the sum of squares; 'df' degrees of freedom (there were 24 observations in all giving 23 total d.f., there were four farms giving 3 d.f. and six fertilizer blends giving 5 d.f.); 'MS' is the mean square (=SS/d.f.) and F is the ratio of the factor MS/error MS; 'P-value' is the important value as it shows the probability that all factor levels (farms or fertilizer blends) have the same mean yield. In this case $P=0.51467$ for 'Rows' (i.e. farms) and $P=0.070826$ for 'Columns' (i.e. fertilizer) so we accept the null hypothesis that farms have the same mean yields and also accept the null hypothesis that the fertilizer blends have the same mean yield. However as the P value was quite close to 0.05 we may consider another trial.

The final value 'F-crit' is not usually quoted on an ANOVA table. It is the value of F required to achieve a $P=0.05$ with the degrees of freedom in this particular test.

ANOVA with more than one observation for each factor combination (with replication)

The situations where you have two factors and more than one observation for each combination of factor levels is a very common one in biology. If there are two factors (ways of dividing the data into classes) then there are three null hypotheses associated with the test:

1   that all levels of the first factor have the same mean;

Chapter 7
The tests 1: tests
to look at
differences

2 that all levels of the second factor have the same mean; and

3 that there is no interaction between the two factors.

If you have no interest in the third null hypothesis then you might consider two separate one-way analyses. However, the extra power of the test that comes from investigation of the interaction is very great. Two tests are considered here: two-way analysis of variance and the Scheirer–Ray–Hare test (page 131).

Interaction

This concept warrants separate consideration. If the test gives a significant result for the interaction term it shows that the effect of the two factors in the test are not additive. That means that groups of observations assigned to levels of factor 1 do not respond in the same way to factor 2. For example, if you are measuring spiders from two locations you could have 'sex' as one factor and 'location' as the other (each spider can be assigned to one combination of sex and location). An analysis of variance gives a significant result for the factor 'sex' and for the factor 'location' as well as the interaction term. This means that in some way the two sexes are responding differently in the two locations. The best way to interpret interaction is to plot out the means of the factor combinations roughly and inspect the graphs (Fig. 7.5).

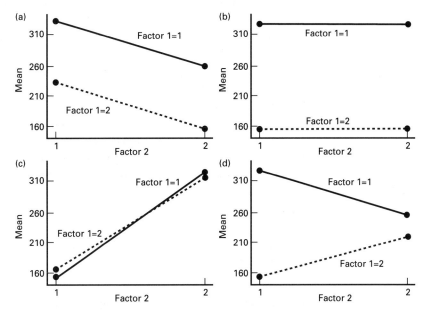

Fig. 7.5 In this figure there are four possible representations of the results of plotting mean values by factors after an analysis of variance. There are two factors labelled 'factor 1' and 'factor 2' here. Both have two levels labelled '1' and '2'. (a) Shows a typical plot where both the main factors are significant but there is no interaction (lines are parallel). In (b) factor 1 is significant, factor 2 is not and there is no interaction. In (c) factor 2 is significant, factor 1 is not and there is no interaction. In (d) both factors are significant and there is also a significant interaction term—the lines are *not parallel* (i.e. the effect of factor 2 is different for the two groups of factor 1).

Chapter 7
The tests 1: tests
to look at
differences

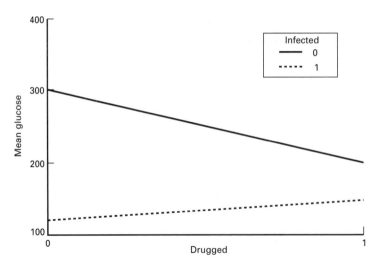

Fig. 7.6 In this example, produced in SPSS, the two lines are clearly not parallel indicating that there is an interaction between the two factors. Here, one factor is having an opposite effect on the two groups defined by the other factor. In the example the effect of the drug depends on whether the individuals are infected or not.

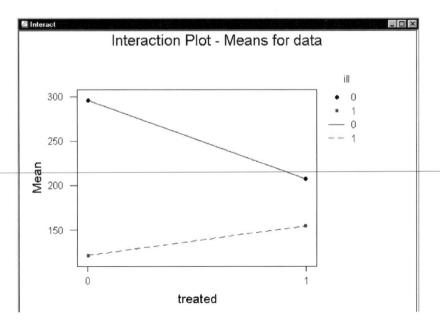

Fig. 7.7 This example uses the default 'interaction plot' option in MINITAB. There were two levels of two factors ('ill' coded 0 or 1 and 'treated' coded 0 or 1). If there were no interaction between the two factors then the line joining the means would be parallel. In this example the lines are clearly not parallel and the treatment is clearly having an opposite effect on subjects (e.g. when 'ill' is 0 there is a decrease in mean level with treatment while when 'ill' is 1 there is an increase).

SPSS

SPSS. Interactions can be visualized in SPSS fairly simply once the analysis (either ANOVA or Scheirer–Ray–Hare) has been carried out. From the 'Graphs' menu select 'Line'. Then in the dialogue box select 'Multiple' and 'Summaries

Chapter 7
The tests 1: tests
to look at
differences

for groups of cases' before clicking on 'Define'. In the next dialogue select 'Other summary function' and move the observations column into the 'Variable:' box (it should default to 'MEAN(column)' if it does not click on 'Change summary' and select mean). Then put the two factors one into the 'Category axis:' and the other into the 'Define lines by:' box (either way is fine, although using the factor with the most group levels as the category axis is usually better). Click 'OK'. The graph produced is shown in Fig. 7.6.

MINITAB

MINITAB. There is a very simple way to visualize interactions in MINITAB. Once the analysis has been carried out following the instructions below then a plot of the means for each factor and the effect of other factors can be produced. From the 'Stat' menu select 'ANOVA' and then 'Interactions plot...' (Fig. 7.7). You can carry out this procedure after either a conventional ANOVA or a Scheirer–Ray–Hare test.

Excel

Excel. Arrange the mean values for each factor combination in a table. Use row and column labels.
     For example:

		Subject ill	
		0	1
Treated	0	295.6667	207.5
	1	121.3333	154.6667

Select these cells. Then either use the 'Chart wizard' or from the 'Insert' menu select 'Chart' and 'On this sheet' before selecting the area where the chart will appear. In step 1 the cells should be selected. In step 2 select 'Line' and then in step 3 select type '1', '2' or '4'. The data should be in 'rows' although often the plot is framed incorrectly making the data appear in 'columns'. There are two rows and columns of labels. In step 5 extra details about the axes can be added. Finally click 'Finished'. A chart something like that shown in Fig. 7.8 appears.

Two-way ANOVA (with replication)

Two-way ANOVA with replication is a very powerful statistical test with three null hypotheses (described above, page 120). It is suitable when there are two independent ways of assigning the observations into groups and there is more than one observation per factor combination. Assumptions are the same as other ANOVA tests: the data are continuous (>30 possible values); at least

[123]

Chapter 7
The tests 1: tests
to look at
differences

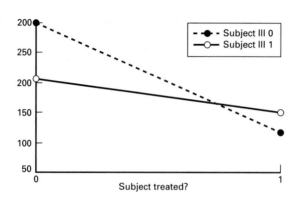

Fig. 7.8 An interaction visualized in Excel. The means have to be calculated first and then plotted. The lines are clearly not parallel indicating that there in an interaction between the two factors. In this example when the subject is not ill ('ill' is 0) there is a decrease in mean level with treatment but when 'ill' is 1 there is an increase in the mean.

approximately normally distributed and the variation is the same in each factor combination. The assumptions appear restrictive but two-way ANOVA is not very sensitive to slight violations of the assumptions. If you have data that clearly doesn't fit the assumptions you have five choices: don't do the test at all; try to transform the data to make it fit the assumptions; carry out a Scheirer–Ray–Hare test in lieu; use two separate one way analyses instead or use the test but be very cautious if the *P* values are anywhere near the critical 0.05 level.

Note: 'Balanced designs' have equal numbers of observations in each factor combination. The statistical calculations are simpler for balanced designs and the tests more powerful. Therefore when you design a controlled experiment it is always sensible to plan for equal numbers of observations from each factor combination (of course this is not always possible when making field collections, etc.). However, as you are not calculating the tests by hand I will not consider 'unbalanced' and 'balanced' designs separately. Just ensure that you avoid factor combinations with no observations at all! Unfortunately Excel will only work with balanced designs for two-way analyses.

An example

A biologist is investigating the effect of light on food intake in starlings. She sets up an experiment where birds are placed in individual aviaries of identical size with controlled lighting and are given an excess of food. Birds are sexed and four male and four female birds are exposed to either 16 hours or 8 hours of light (e.g. long or short days). Each bird is monitored for 7 days and its total food intake in grams recorded. Each bird is only used once.

Chapter 7
The tests 1: tests
to look at
differences

Day-length	Sex							
	Female				Male			
Long (16 h)	78.1	75.5	76.3	81.2	69.5	72.1	73.2	71.1
Short (8 h)	82.4	80.9	83.0	88.2	72.3	73.3	70.0	72.9

There are sixteen birds used in total. Four were exposed to each factor combination. This is a balanced design.

The output from the statistical packages varies somewhat. It is usually not sensible to just put the whole output into a report. For a two-way ANOVA there is a standard way of displaying the output from the test. For the example here the output would be represented in the following way:

Factor	d.f.	SS	MS	*F* ratio	*P* value
Day length	1	42.3	42.3	8.00	0.015
Sex	1	316.8	316.8	60.00	<0.001
Interaction	1	27.0	27.0	5.12	0.043
Error	12	63.4	5.28		

SPSS

SPSS. Input all the observations in a single column. Use two further columns to input codes for the group labels associated with the two factors. The factor levels must be coded as integers. Label the columns appropriately. There are then two methods you can use to carry out the test that give somewhat different output. Method 2 has more scope for requesting additional information, summaries of the data etc.

Method 1. From the 'Statistics' menu select 'ANOVA models' then 'Simple factorial...'. In the dialogue box move the observation column into the 'Dependent:' box and the two coded factor columns into the 'Factor(s):' box. You must then tell SPSS what the code numbers for the lowest and highest factor levels are. Select the factors in turn, click on the 'Define Range...' range and put the appropriate numbers into the 'Minimum' and 'Maximum' boxes. (If both factors have the same range of factor levels then you can select both before defining the range.) Click 'continue' to return to the main box and then 'OK' to run the test.

Using the example you would get the following output:

Chapter 7
The tests 1: tests
to look at
differences

```
* * *  A N A L Y S I S   O F   V A R I A N C E  * * *

                    FOOD_G
            by      DAY_LENG
                    SEX

                UNIQUE sums of squares
                All effects entered simultaneously


                                  Sum of                 Mean            Sig
Source of Variation               Squares      DF       Square      F    of F

Main Effects                      359.090       2      179.545  33.999   .000
    DAY_LENG                       42.250       1       42.250   8.001   .015
    SEX                           316.840       1      316.840  59.998   .000

2-Way Interactions                 27.040       1       27.040   5.120   .043
    DAY_LENG SEX                   27.040       1       27.040   5.120   .043

Explained                         386.130       3      128.710  24.373   .000

Residual                           63.370      12        5.281

Total                             449.500      15       29.967


16 cases were processed.
0 cases (.0 pct) were missing.
```

Data output 7.43

The output confirms the test used, then the dependent variable name (the observations) and the names of the factors used. The next two lines are relevant to the way the calculation was carried out in the package. Then comes the ANOVA table itself. If you compare the output with the table I used in the example you will see that there are extra lines. These are amalgamations of other lines and are not really very useful. The lines you should look at are those labelled with the factors ('day_leng' and 'sex' in the example), that for the interaction of the two factors (labelled as 'day_leng sex' here) and that for the error or residual as it is labelled here. The five columns in the table give first the 'sum of squares', then the 'DF' column gives the degrees of freedom, which is one less than the number of factor levels. The mean squares are given (mean square = sum of squares/degrees of freedom). The last two columns give the important information. The 'F' column is the F ratio (factor mean square/residual mean square). The P-value is labelled 'Sig of F' and given in the last column. There are three null hypotheses to look at: levels of factor 1 have the same mean; levels of factor 2 have the same mean; there is no interaction between factor 1 and factor 2. If any of the P values are less than 0.05 then you must reject the relevant null hypothesis.

In the example the P value for 'sex' is given as '0.000'; this is a highly significant result but should always be reported as $P < 0.001$. We can be sure that the two sexes have different food intake levels although the test does not tell us which sex eats more — that must be done by inspection of the data. The P value for the other factor 'day_leng' is also less than 0.05 but the value of 0.015 indicates that the effect is not as strong as that for sex. Finally the interaction of the two factors also has a P value less (albeit slightly less) than 0.05 meaning, in this example, that the two sexes respond significantly differently to day length in the amount of food they eat. The direction of this

Chapter 7
The tests 1: tests
to look at
differences

effect can only be revealed by inspection of the mean values for each group (see the section on interaction).

Method 2. From the 'Statistics' menu select 'ANOVA models' then 'General factorial…'. In the dialogue box move the observation column into the 'Dependent:' box and the two coded factor columns into the 'Factor(s):' box. Select the factors in turn, click on the 'Define Range…' range and put the appropriate numbers into the 'Minimum' and 'Maximum' boxes (Fig. 7.9, page 129). Click 'continue' to return to the main box and then 'OK' to run the test. Note: if you click on 'Options…' button before running the test you can ask for summary information about the various factor levels. This extra information can be useful when interpreting the results of the test.

You get the following minimum output:

```
* * * * * * A n a l y s i s   o f   V a r i a n c e -- design  1 * * * * * *

Tests of Significance for FOOD_G using UNIQUE sums of squares
Source of Variation          SS        DF       MS         F   Sig of F

WITHIN+RESIDUAL            63.37       12      5.28
DAY_LENG                   42.25        1     42.25      8.00     .015
SEX                       316.84        1    316.84     60.00     .000
DAY_LENG BY SEX            27.04        1     27.04      5.12     .043

(Model)                   386.13        3    128.71     24.37     .000
(Total)                   449.50       15     29.97

R-Squared =        .859
Adjusted R-Squared =  .824
```

Data output 7.44

This confirms the test used and gives an indication of the name of the variable where the observations are. The table has the usual ANOVA table headings: 'SS' is sum of squares; 'DF' is degrees of freedom; 'MS' is mean square; 'F' is the *F* ratio and 'Sig of F' is the *P* value. The two coded factors have a row each ('day_leng' and 'sex' in the example). The row labelled 'within + residual' is the error or residual variation while 'day_leng by sex' refers to the interaction. The extra lines of '(model)', '(total)' and 'R-squared' show how much of the variation in all the observations is accounted for by the differences in the factor levels. For interpretation of the *P*-values see method 1 above.

MINITAB

MINITAB. Input all the observations into a single column. Use separate columns to input the factor levels, coded as integers. Label the columns appropriately. There are then at least two ways of carrying out the test that give rather different outputs. I will only consider one method here.

From the 'Stat' menu, select 'ANOVA' then 'General linear model…'. In the dialogue box put the column containing the data into the 'Responses:' box and then the other two factors into the 'Model:' box. To investigate interaction

Chapter 7
The tests 1: tests
to look at
differences

between the two factors the two column labels should be moved into the 'Model:' box again and an asterisk (*) put between them (in the example the 'Model:' box contained: sex 'day leng' sex * 'day leng'). Before continuing it is useful to click on 'Options…' and put the factor variables into the 'Display means for…' box. This generates some additional output useful when interpreting the results of the test. Click 'OK' in this box and again in the first dialogue box. (Or type 'glm c1 = c2 c3 c2*c3;' at the MTB> prompt and 'means c2 c3.' at the SUBC> prompt.)

Using the example (and selecting 'display means for') the following output appears:

```
General Linear Model

Factor     Levels Values
sex           2    1    2
day leng      2    1    2

Analysis of Variance for food g

Source         DF    Seq SS    Adj SS    Adj MS      F      P
sex             1    316.84    316.84    316.84    45.56  0.000
day leng        1     42.25     42.25     42.25     6.08  0.028
sex*day leng    1     27.04     27.04     27.04     5.12  0.043
Error          12     63.37     63.37      5.28
Total          15    449.50

Unusual Observations for food g

Obs.    food g       Fit Stdev.Fit  Residual   St.Resid
 12     88.2000   82.3250   1.1419    5.8750      2.47R

R denotes an obs. with a large st. resid.


Means for food g

sex           Mean     Stdev
 1           80.70    0.9324
 2           71.80    0.9324
day leng
 1           74.62    0.9324
 2           77.88    0.9324
```

Data output 7.45

First comes confirmation of the test (note that 'General linear model' includes ANOVA) and then the factors used, the number of levels (groups) and the integer codes assigned to the groups. (In the example there are obviously two levels for the factor 'sex' and they have been coded alphabetically as '1' for female and '2' for male.)

The standard ANOVA table is next. It labels the factors and the interaction. The columns give degrees of freedom 'DF'; two versions of sums of squares 'Seq SS' and 'Adj SS'; the mean square 'Adj MS' which is SS/df; the F ratio 'F' which is MS/MS_{error} and finally the P value 'P'.

In the example the factor 'sex' has a P value of 0.000 which should be reported as $P < 0.001$. We reject the null hypothesis that the two sexes have the same food intake confidently although we can't tell from the table which sex has the greater intake. The factor 'day leng' and the interaction also have P values less than 0.05 so we reject the null hypotheses that the factor levels have the same mean food intake but note that the result is not highly significant.

Chapter 7
The tests 1: tests
to look at
differences

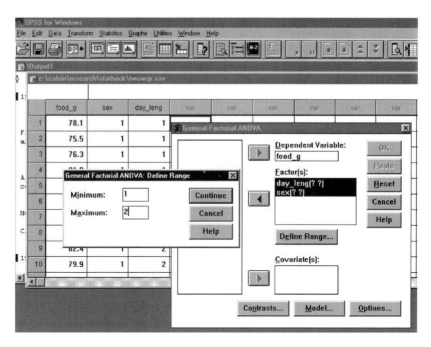

Fig. 7.9 Two-way analysis of variance in SPSS. The two factors have been highlighted and the 'Define range...' button has been selected. Both factors have levels coded '1' and '2' and this is being applied to both simultaneously.

The next part of the output gives a list of unusual observations; it is always wise to check that these have been input correctly into the package. Finally, the additional output requested on the means is shown. This allows interpretation of the significant results in the ANOVA table. In the example sex group '1' (females) have higher food intake and day length group '2' (short days).

Excel

Excel. This test is only possible in Excel if the design is balanced (i.e. each factor combination has the same number of observations). The data must be input so that one of the factors is separated into different columns and the other in sets of rows. The columns must be labelled with group labels and the first row of each group likewise.

The example was set up like this:

	Long	Short
Female	78.1	82.4
	75.5	80.9
	76.3	83
	81.2	88.2

Continued on p. 130

Chapter 7
The tests 1: tests
to look at
differences

	Long	Short
Male	69.5	72.3
	72.1	73.3
	73.2	70
	71.1	72.9

From the 'Tools' menu select 'Data analysis…' then 'ANOVA: two factor with replication'. In the dialogue box that appears you must select the range of cells that contains not only the data but also the labels. This can be done either by dragging the mouse over the area of the relevant section of the spreadsheet or typing the top left and bottom right cell codes with a colon ':' between them. Next, state how many observations there were for each factor combination. Leave the 'Alpha:' as 0.05 (this determines the critical F value which is quoted on the final table for comparison, i.e. the F value required to reject the null hypothesis).

Finally, to select where the output will appear select one of the options in the bottom half of the box. Beware—the output covers a lot of cells.

The output is rather extensive:

```
Anova: Two-Factor With Replication

SUMMARY    Long        Short     Total
   Female
Count            4          4           8
Sum          311.1      334.5       645.6
Average     77.775     83.625       161.4
Variance  6.395833    10.0825    16.47833


   Male
Count            4          4           8
Sum          285.9      288.5       574.4
Average     71.475     72.125       143.6
Variance  2.469167   2.175833       4.645


   Total
Count            8          8
Sum            597        623
Average     149.25     155.75
Variance     8.865   12.25833


ANOVA
Source of     SS         df         MS          F         P-value      F crit
Variation
Sample      316.84        1       316.84     59.99811    5.22E-06 4.747221
Columns      42.25        1        42.25      8.000631    0.015217 4.747221
Interacti    27.04        1        27.04      5.120404    0.042991 4.747221
on
Within       63.37       12      5.280833

Total        449.5       15
```

Data output 7.46

The first part of the output confirms the test carried out and then presents a summary table giving number of observations, sum, mean and variance for

Chapter 7
The tests 1: tests
to look at
differences

each factor combination, factor level and total. Some of this information is useful, especially when trying to interpret a significant result from the interaction term.

Then comes a rather standard ANOVA table that will require minimum editing before it can be included in a report. If you compare it with the 'ideal' table in the example you see how similar it is. The two factors are, unfortunately, only assigned as 'columns' and 'sample' although inspection of the original data in Excel will quickly determine which factor is which. The residual or error line is labelled as 'within'. The columns of figures are sum of squares (SS), degrees of freedom (d.f.), mean square (MS, where $MS = SS/d.f.$), F ratio (F, where $F = MS/MS_{error}$) the P value and then the F ratio that was required to reach a P value of 0.05. The P values are sometimes given in scientific notation, so the '5.22E-06' in the example translates as 5.22×10^{-6} or 0.0000522 – extremely highly significant!

In the example all three P values are less than 0.05 so all three original null hypotheses are rejected.

Scheirer–Ray–Hare test

Nonparametric two-way ANOVA is usually deemed impossible in statistics books but the Sheirer–Ray–Hare test is a nonparametric equivalent of a two-way ANOVA with replication. It is based on ranks so is suitable for any situation where the data can be put into order. It is quite easy (especially with fairly small data sets) to calculate a two way ANOVA using ranks in an extension of the Kruskal–Wallis test called the Scheirer–Ray–Hare test. It makes few assumptions about the distribution of the data. However, the test is fairly recent and not yet widely available in statistical packages. For that reason it is a little awkward to carry out. This test is conservative and has much lower power than the parametric ANOVA. There are further extensions to this test just becoming available to allow testing of multiway ANOVA.

Example

Using the same data of starlings grouped by sex and day length as for the two-way ANOVA. In contrast to the conclusions from the standard ANOVA the Sheirer–Ray–Hare test accepts two of the three null hypotheses associated with the experiment: there is no interaction between the two factors, and birds in different day lengths consume the same amount of food.

SPSS

| SPSS

1 Set out the data in exactly the same way as for parametric ANOVA (page 125). Sort the data (ascending) using the magnitude of the observations. From the 'Data' menu select 'Sort cases…'. In the dialogue box move the data column name into the 'Sort by:' box and select 'Ascending' (although it does not matter whether you choose ascending or descending). Click 'OK' and the

Chapter 7
The tests 1: tests
to look at
differences

rows will be shuffled into rank order.

2 Next you should assign ranks to replace the actual data (assign 1 to the smallest etc.). You can either write over the original data observations or make a new column called 'rank'. Type the row number into this column.

3 Carry out a standard (parametric) ANOVA including the interaction. From the 'Statistics' menu select 'ANOVA models' then 'Simple factorial'. In the dialogue box put the 'rank' column into the 'Dependent:' box and the two coded factor columns into the 'Factor(s)' box. Define the range of the factor levels (as in two-way ANOVA). Click 'OK' to run the test. You will get this output if you use the example:

```
* * *   A N A L Y S I S   O F   V A R I A N C E   * * *

            RANK
    by   DAY_LENG
         SEX
                            Sum of                Mean               Sig
Source of Variation         Squares      DF       Square      F      of F

Main Effects                281.000       2       140.500   30.655   .000
    DAY_LENG                 25.000        1       25.000     5.455   .038
    SEX                      256.000       1       256.000   55.855   .000

2-Way Interactions          4.000         1       4.000      .873    .369
    DAY_LENG SEX            4.000         1       4.000      .873    .369

Explained                   285.000       3       95.000    20.727   .000

Residual                    55.000       12       4.583

Total                       340.000      15       22.667

16 cases were processed.
0 cases (.0 pct) were missing.
```

Data output 7.47

---

4 This is not the end point for this test. You have to do the next bit by hand. Note down the Mean Square value in the 'Total' row, call this value MS_{total}. (In the example MS_{total} is 22.667.)

The test statistic for the three null hypotheses is the value for relevant Sum of Squares (SS) divided by MS_{total}. Calculate SS/MS_{total} for each factor and the interaction.

Using two new columns in SPSS type in the values of SS/MS_{total} and degrees of freedom using the same degrees of freedom as in a conventional ANOVA (i.e. one less than the number of factor levels for the two main factors and the degrees of freedom for the main factors multiplied together for the interaction). Label the columns, 'out' and 'df'. Then, to get the appropriate *P* values for these values, from the 'Transform' menu select 'Compute…'. In the 'compute variable' window that appears type the name of the column you want the result to go to (e.g. 'pvalue'). Then in the 'Numeric expression:' box type '1–' (i.e. one minus) before selecting from the 'Functions' list 'CDF.CHISQ(q,df)' and moving it into the 'Numeric expression:' box. Finally, replace the 'q' with the name of the variable where you input the SS/MS_{total} values and the 'df' with the name of the column with the degrees of freedom.

Chapter 7
The tests 1: tests
to look at
differences

Click 'OK' and the new column 'pvalue' will appear on the right-hand side of the spreadsheet with P values from a chi-square table appropriate to the SS/MS_{total} values and degrees of freedom. (Note you could look up the number in a chi-square table if you prefer.)

So, for the example (using the same labels as the SPSS output):

	SS	SS/MS_{total}	d.f.	*P* value
Day_leng (Factor)	25	1.10	1	0.29427
Sex (Factor)	256	11.29	1	0.00078
Day_leng Sex (Interaction)	4	0.176	1	0.67483

The results suggest that the only null hypothesis that must be rejected is that both sexes consume the same amount of food as the P value is <0.001.

MINITAB

MINITAB. There is no way of carrying out this test directly in MINITAB but with a little extra work the test is achievable.

1 Put the observations into a single column then the factors in the next two columns with the factor levels coded as integers. Label the three columns appropriately. From the 'Manip' menu select 'Rank...'. Move the data observation column into the 'Rank data in:' box and in the 'Store ranks in:' type a new name (e.g. 'rank'). Click 'OK' and the rank position of the data appears in the new column with the smallest value ranked as 1.

2 Carry out a conventional two-way ANOVA using the 'rank' rather than the original observations. From the 'Stat' menu select 'ANOVA' then 'Twoway...'. Put the rank column into the 'Response:' box and the two factors into the 'Row factor:' and 'Column factor:' boxes. Click 'OK'. (Or, assuming the ranks are in c4 and the factors in c2 and c3, type 'twow c4 c2 c3' at the MTB> prompt in the session window.) The example will give the following output:

```
Two-way Analysis of Variance

Analysis of Variance for rank
Source          DF        SS        MS
sex              1    256.00    256.00
day leng         1     25.00     25.00
Interaction      1      4.00      4.00
Error           12     55.00      4.58
Total           15    340.00
```

Data output 7.48

3 Record the SS (sum of squares) values for the two factors and the

Chapter 7
The tests 1: tests
to look at
differences

interaction and calculate the total MS (SS_{total}/DF). Then calculate values for SS/MS_{total} either in MINITAB using 'Calc', then 'Numerical expression...' or by using a calculator. Type the calculated values into a single column in the spreadsheet (make sure to remember which value applies to each factor). Then use MINITAB to look up the values on a chi-square table: from the 'Calc' menu select 'Probability distributions' then 'Chi-square...'. Select the 'cumulative probability' option. Then input the appropriate degrees of freedom (if the factors and interaction have different degrees of freedom then you will have to repeat this process for each different number of degrees of freedom); in the example there is one degree of freedom for all three values. Then in the 'Input column:' box put the name of the variable containing the calculated values and choose an empty column for the output 'Optional storage:'. Click 'OK'. This generates values for the probability of getting the number calculated *or lower* whereas the number *or higher* is required for a *P* value. So from the 'Calc' menu select 'Numerical expression...'. In the dialogue window put the name 'pvalue' in the 'Variable (new or modified):' box. Finally in the 'Expression:' box type '1–' then the name of the column where the chi-square output was sent.

Using the example the table below shows the numbers used.

	SS	MS	SS/MS$_{total}$	d.f.	Cumulative chi-square value	*P* value
Day leng (factor)	25		1.10	1	0.705734	0.29427
Sex (factor)	256		11.29	1	0.999221	0.00078
Interaction	4		0.176	1	0.325166	0.67483
Total	340	22.67		15		

In the example the *P* values for 'day length' and 'interaction' are both well above 0.05 so the null hypothesis that there is no difference between factor levels is accepted. However, the *P* value for 'sex' is <0.001 indicating that the null hypothesis should be rejected and the two sexes have highly significantly different food intake.

Excel

Excel. There is no direct way of carrying out this test in Excel although, providing the design is balanced (there is the same number of observations in each factor combination), it can be done using two-way ANOVA with only a few extra steps to rank the data and look up the significance of the test result.
1   First the raw data should be turned into ranks starting with the lowest value as 1 and then put into a table with row and column headings. The example data set becomes:

[134]

Chapter 7
The tests 1: tests
to look at
differences

	Female	Male
Long	9	1
	10	3
	11	4
	13	7
Short	12	2
	14	5
	15	6
	16	8

2   Carry out a 'Two-way ANOVA with replication' exactly as described above. From the 'Tools' menu select 'Data analysis…' and then 'ANOVA: two-factor with replication'. Define the range to cover the cells with the ranked data and the group labels. Indicate how many observations there are for each factor combination, leave 'Alpha' as 0.05 and send the output to a clear area of the spreadsheet. Click 'OK'.

3   Ignore the output except for the ANOVA table at the bottom. Cut out the first four columns of the output and paste it into the top left corner of a new sheet. For the example the required output is:

Source of Variation	SS	df	MS
Sample	25	1	25
Columns	256	1	256
Interaction	4	1	4
Within	55	12	4.583333
Total	340	15	

Data output 7.49

Moving the output like this is not required by the package but it does remove distractions!

4   Calculate the total mean square MS_{total}. This is the total sum of squares (SS) divided by the total degrees of freedom (d.f.). If you have moved the table so that cell 'a1' contains the text 'Source of Variation' then just type in cell d7: '=b7/c7'. The statistic for the Scheirer–Ray–Hare test can now

Chapter 7
The tests 1: tests
to look at
differences

be calculated for the two main factors and the interaction as it is SS/MS$_{total}$. In column 'e' type '=b2/d7' in row 2, '=b3/d7' in row 3 and '=b4/d7' in row 4. This gives three numbers that need to be looked up on a chi-square table — very easily done in Excel.

5   In column 'f' type 'P-value' in row 1 as a label. Then in row 2 type '=chidist(e2,c2)' and the P value for the value in column 'e' and the degrees of freedom in column 'c' is given. Copy this into the two cells below (easily done in Excel by selecting cell 'f2' then clicking on the small black square in the bottom right corner of the cell and dragging it down to cover the next two cells).

The final table for the example data is as follows:

Source of Variation	SS	df	MS	P-value
Sample	25	1	25 1.102941	0.293622
Columns	256	1	256 11.29412	0.000778
Interaction	4	1	4 0.176471	0.674424
Within	55	12	4.583333	
Total	340	15	22.66667	

Data output 7.50

The P values for 'Sample' (= day length in the example) and 'Interaction' are well above 0.05 so the null hypothesis that the factor levels have the same median is accepted. The P value for 'Columns' (= sex) is well below 0.05. The null hypothesis is rejected and the alternative hypothesis that the two groups (sexes) have different food intakes is accepted. Which level is the higher must be determined by inspection.

There are more than two independent ways to classify the data

If each observation can be assigned to groups using more than two different factors (ways of classifying the data), each of the factors can be divided into groups and assigned integers and, most importantly, the factors are all fully independent of each other then the data is suitable for multifactorial testing.

The biggest danger is thinking that factors are independent when they are not. For example if you sample from a variety of woodlands and use wood number as a factor then divide the samples into two groups based on whether they were in a northern or southern area. The woodland number is not independent of the region (e.g. wood number 6 can only be in one or the other). If this is the case then a nested design is required rather than a 'fully factorial' one (see page 142).

Chapter 7
The tests 1: tests
to look at
differences

Multifactorial testing

If there are more than two factors and they are fully independent of each other then the design is said to be factorial. When there were just two factors there was scope for only one interaction term: factor A with factor B. As more factors are added to the analysis the number of interaction terms possible grows rapidly. For three factors there are four interaction terms (A×B, A×C, B×C and A×B×C), for four factors there are eleven.

Three-way ANOVA (without replication)

When time, space or resources are very limited then this design is quite common. If there are three independent ways of classifying the data into groups and there is only one observation for each factor combination then it is still possible to carry out an analysis of variance. Follow the instructions for three-way ANOVA with replication but be aware that there is no way of calculating the interaction between the three factors (this is used as the 'error' by the calculation). This means that only two-way interactions should be allowed in the model.

Three-way ANOVA (with replication)

If there are three fully independent ways of dividing the data into groups (e.g. site of collection, species and sex) and there is more than one observation for each factor combination then this design is appropriate. As with all analysis of variance the test assumes that the data is continuous, approximately normally distributed and that the variance is approximately equal for each factor level. Unfortunately, there is no nonparametric equivalent to use if the assumptions are not met.

An example

A group of agricultural ecologists are interested in the effect of rabbit grazing on plant communities near to warrens. The observations are carried out on two sites on different soils. At each site there are two areas for study, one is 10 m from an active warren and one is 100 m from a warren. In each study area six quadrats are positioned at random. Two quadrats are left untouched as a control, two have caged exclosures placed around them that totally exclude rabbits and two are 'procedural controls' where the digging for constructing an exclosure is carried out but the rabbits are not excluded. Required in this experiment as otherwise any effects could be attributed either to the disturbance from construction or the exclusion of rabbits. Procedural controls are often overlooked when designing experiments (see page 26 for a description of procedural control).

  The data collected is the standing crop of grass collected 100 days after the exclosures were placed. There are three independent factors: site (two levels coded 1 and 2); distance from warren (two distances 10 m and 100 m,

Chapter 7
The tests 1: tests
to look at
differences

coded 1 and 2); exclosure type (three levels coded: 0, control; 1, procedural control; 2, exclosure). There are two observations for each factor combination for a total of 24.

Here is the data in table format:

		Distance from warren					
		1			2		
		Exclosure type					
		0	1	2	0	1	2
Site	1	112	115	187	141	121	189
		116	102	175	101	157	186
	2	121	145	198	135	141	208
		138	124	168	129	133	206

Output from statistical packages often has more information than is required. The ANOVA table included in a report or publication should be presented as follows:

Source of Variation	d.f.	SS	MS	*F*	*P*
Distance	1	1218.4	1218.4	6.44	0.026*
Exclosure	2	6841.4	3420.7	18.07	<0.001***
Site	1	590.0	590.0	3.12	0.103
Distance × Exclosure	2	327.0	163.5	0.864	0.446
Distance × Site	1	1.0	1	0.006	0.942
Exclosure × Site	2	58.3	29.2	0.154	0.859
Distance × Exclosure × Site	2	282.3	141.2	0.746	0.495
Error	12	2271.5	189.3		
Total	23				

This sort of table should always be accompanied with a table caption that explains the test used in more detail without making the reader refer to the main text.

Chapter 7
The tests 1: tests
to look at
differences

SPSS

SPSS. There are at least two methods using SPSS. They are essentially the same as for two-way ANOVA. Arrange the data so that there is one column for the observations and one for each of the three factors with the group labels coded as integers.

Method 1. From the 'Statistics' menu select 'ANOVA models' then 'Simple factorial...'. Put the observations into the 'Dependent:' box and the three factors into the 'Factor(s):' box. Define the range (highest and lowest coded integer) for each factor. Click 'OK' to run the test.

(*Note*: if you have a three way ANOVA without replication (i.e. there is only one observation for each factor combination) you must first go to the 'Options...' and restrict the model to 'two-way interactions only'.)

Using the example you get the following output:

```
* * *  A N A L Y S I S   O F   V A R I A N C E  * * *

                GRASS_WT
         by     DISTANCE
                EXCLOSE
                SITE

              UNIQUE sums of squares
              All effects entered simultaneously

                              Sum of              Mean               Sig
Source of Variation           Squares    DF       Square     F       of F

Main Effects                  8649.750    4       2162.438   11.424  .000
    DISTANCE                  1218.375    1       1218.375    6.436  .026
    EXCLOSE                   6841.333    2       3420.667   18.071  .000
    SITE                       590.042    1        590.042    3.117  .103

2-Way Interactions             386.375    5         77.275     .408  .834
    DISTANCE EXCLOSE           327.000    2        163.500     .864  .446
    DISTANCE SITE                1.042    1          1.042     .006  .942
    EXCLOSE  SITE               58.333    2         29.167     .154  .859

3-Way Interactions             282.333    2        141.167     .746  .495
    DISTANCE EXCLOSE  SITE      282.333    2        141.167     .746  .495

Explained                     9318.458   11        847.133    4.475  .008

Residual                      2271.500   12        189.292

Total                        11589.958   23        503.911

24 cases were processed.
0 cases (.0 pct) were missing.
```

Data output 7.51

The output confirms the test and then gives the name of the observation column and then the factor names. The next two lines show that the ANOVA is using standard assumptions. Then comes the ANOVA table proper. There is more information than you actually need here. The most important lines are for the factor variables, called 'Main Effects' in SPSS ('Distance', 'Exclose' and 'Site' in the example) and their interactions (three possible two-way interactions and one three way). The 'DF' column gives the degrees of freedom which is one less than the number of factor levels for main effects and calculated by multiplication for interactions (e.g. for 'distance exclose' there are $2 \times 1 = 2$ d.f.). The sum of squares and mean square are given,

Chapter 7
The tests 1: tests
to look at
differences

mean square is sum of squares/degrees of freedom. The last two columns give the important information. The 'F' column is the F ratio (factor mean square/residual mean square). The P value, labelled 'Sig of F', is in the last column. In the example the P value for 'Site' is slightly above the critical 0.05 level so we accept the null hypothesis that there is no difference between sites. However, the P values for 'Distance' and 'Exclose' are both less than 0.05 so we reject the null hypotheses and accept the alternative hypotheses that distance from warren and the presence of exclosures affect the grazing of rabbits. None of the interaction terms has a P value even close to 0.05.

In this output it is best to totally ignore the lines labelled 'Main Effects', '2-way Interactions' and '3-way Interactions' and 'Explained' as they are produced by combining values.

Method 2. From the 'Statistics' menu, select 'ANOVA models' then 'General factorial...'. Put the observations into the 'Dependent variable:' box and the three factors into the 'Factor(s):' box. Define the ranges of each of the factors (if you are unsure how to do this see the section on one-way ANOVA on page 86), then click 'OK'.

This is the output from the example data:

```
* * * * * * A n a l y s i s   o f   V a r i a n c e -- design   1 * * * * * *

Tests of Significance for GRASS_WT using UNIQUE sums of squares
Source of Variation          SS      DF      MS        F   Sig of F

WITHIN+RESIDUAL            2271.50   12    189.29
DISTANCE                   1218.38    1   1218.38     6.44    .026
EXCLOSE                    6841.33    2   3420.67    18.07    .000
SITE                        590.04    1    590.04     3.12    .103
DISTANCE BY EXCLOSE         327.00    2    163.50      .86    .446
DISTANCE BY SITE              1.04    1      1.04      .01    .942
EXCLOSE BY SITE              58.33    2     29.17      .15    .859
DISTANCE BY EXCLOSE         282.33    2    141.17      .75    .495
 BY SITE

(Model)                    9318.46   11    847.13     4.48    .008
(Total)                   11589.96   23    503.91

R-Squared =      .804
Adjusted R-Squared =   .624
```

Data output 7.52

The error term is given first, then the three factors, the three possible two-way interactions, the one three-way interaction and finally some summary lines. In the example only the main factors 'distance' and 'exclose' proved significant. For a more detailed consideration of the output see method 1 or two-way ANOVA with replication.

MINITAB

MINITAB. Put the observations into a single column. Use three further columns for the group labels of the three factors (coded as integers). Label the columns appropriately. From the 'Stat' menu select 'ANOVA' then 'General Linear Model...'. Move the observation column into the 'Responses:' box and

Chapter 7
The tests 1: tests
to look at
differences

the three factor columns into the 'Model:' box. Unfortunately if interactions are to be investigated these need to be added separately. So, assuming the factors are in columns 2, 3 and 4 then the following needs to be added to the model box: 'c2*c3 c2*c4 c3*c4 c2*c3*c4'. The names of the factors could be used instead of column codes. [If there is no replication (i.e. only one observation for each factor combination) then the final c2*c3*c4 term should be omitted.]

The following output appears:

```
General Linear Model

Factor    Levels Values
site        2    1    2
distance    2    1    2
exclose     3    0    1    2


Analysis of Variance for grass wt

Source                  DF    Seq SS    Adj SS    Adj MS      F       P
site                     1     590.0     590.0     590.0    3.12   0.103
distance                 1    1218.4    1218.4    1218.4    6.44   0.026
exclose                  2    6841.3    6841.3    3420.7   18.07   0.000
site*distance            1       1.0       1.0       1.0    0.01   0.942
site*exclose             2      58.3      58.3      29.2    0.15   0.859
distance*exclose         2     327.0     327.0     163.5    0.86   0.446
site*distance*exclose    2     282.3     282.3     141.2    0.75   0.495
Error                   12    2271.5    2271.5     189.3
Total                   23   11590.0

Unusual Observations for grass wt

Obs. grass wt      Fit Stdev.Fit  Residual   St.Resid
  5   141.000  121.000    9.729    20.000      2.06R
  6   101.000  121.000    9.729   -20.000     -2.06R

R denotes an obs. with a large st. resid.
```

Data output 7.53

This is essentially the same as the output for two-way ANOVA only with several more lines. I always advise checking the first part of the output confirming the factor names, number of groups and the integer codes. The ANOVA table contains lines for each of the factors, then the interactions and finally the error and total. The P values (labelled 'P') are less than 0.05 for only two of the factors and none of the interactions. The null hypotheses that grass weight is the same for all levels of these two factors is rejected and an alternative hypothesis is accepted.

Visualization of which factor level is higher than which is simple to achieve in MINITAB. From the 'Stat' menu select 'ANOVA' then 'Interactions plot…'. Put the factor columns into the 'Factors:' box (no need for the interaction terms this time) and the raw data in the 'Raw response data in:' box. Click 'OK' and a set of graphs showing the mean values of the raw data for every factor combination appears. (See interaction section for an explanation of how these graphs can be interpreted, page 121).

Excel

Excel. There is no direct way to carry out this test using Excel.

Chapter 7
The tests 1: tests
to look at
differences

Multiway ANOVA

If there are more than three ways of dividing the data into groups and each of the classifications is independent of the others then ANOVA may be carried out. These multifactorial designs become increasingly difficult to interpret because there is an explosive increase in the number of interaction terms as the number of factors goes up. Furthermore, unless an experiment is designed to be fully factorial there are likely to be combinations of factors where there are no observations.

Finally, once there are many factors the chance that they are all independent of each other diminishes. Each factor should be considered in turn — is it independent of all other factors? Can two factors be combined in some way? Is one factor nested within another (see next section)? Is one factor more appropriately investigated as a covariate (see section on analysis of covariance, page 172)?

Not all classifications are independent

It is important to identify ways of classifying the data as nonindependent of each other and once that is achieved the correct way to proceed with analysis is clear. A very large number of experiments that are treated as if all the factors are independent prove to be, on further investigation, nothing of the sort.

Nonindependent factors. Some ways of classifying data are always going to be fixed, and therefore, main factors in an analysis. For example, a commonly used factor in biology is 'sex' there are two sexes (although in some species there are intermediate 'intersexes' or hermaphrodites) and once an individual is labelled as either male or female that classification is fixed. Other factors are not so clearly fixed: if a set of observations is divided into two equal-sized groups by weight then an individual may find itself in the heavy group at the start but later moved into the light group. This type of factor is not really fixed although the label 'heavy' really means something about the observation. Consider a study on blood pressure in humans using the two factors I have described here: 'sex' and 'weight'. Males are heavier than females and therefore there will be more males in the heavy group; therefore the two factors are not independent of each other. This type of independence can be checked using a chi-square test of association considered in the next chapter (page 147). If the factors are not independent then two-way ANOVA is inappropriate. One solution is to apply an analysis of covariance (ANCOVA) using 'sex' as a fixed factor and 'weight' as a covariate. This analysis is considered in the next chapter (page 172).

Nested factors. These are never 'fixed' meaningful factors, they are always factors numbered for convenience where the number given does not really mean anything. For example in a greenhouse experiment there may be plants in numbered pots with four pots per numbered tray and three trays in each of

Chapter 7
The tests 1: tests
to look at
differences

two greenhouses. At each level 'pot', 'tray' and 'greenhouse' the numbers are only used for convenience. These classes of factors are called 'random' factors. In this example the 'pots' are said to be nested within 'trays' and the trays within 'greenhouse'. It is important that the numbering system reflects this nesting: pots should be labelled 1–4 in each tray and trays labelled 1–3 in each greenhouse. *Nested factors are always random but not all random factors need to be nested.*

Random or fixed factors. I have touched on the difference between 'fixed' or meaningful factors and 'random' or convenience factors above. There is, however, not a strict distinction between the two types of factors and the same factor can be treated as 'fixed' or 'random' in different tests. For example in an analysis of behaviour in *Drosophila melanogaster* mutants the mutants used could be either a 'fixed' or 'random' factor. If the mutant is treated as 'fixed' (so that *vestigial* is coded as 1 and *white-eye* as 2) then this implies that the significance level of any difference is a result of the characteristics of the particular named mutants. However, if the factor is 'random' then any significance implies differences between any two randomly selected mutants.

This section considers designs with a simple hierarchy of nested factors. More complicated designs where it is appropriate to test both nested and factorial designs are combined are surprisingly common. Despite the fact that these designs are extremely common in biology they are difficult to achieve in the statistical packages.

Nested or hierarchical designs

If the experiment or sampling design has only one 'random' factor at the top of the hierarchy and all other factors are 'nested' within it then the design is said to be a pure nested one. If the top level is 'fixed' and there are other factors nested within it the design is often called a 'mixed model'. There is no limit to the number of factors that can be nested in this way. The simplest case of a single fixed factor and one nested is considered first then a more complex one.

Two-level nested design ANOVA

Two-level nested design ANOVA is an analysis with two factors. The standard factor in this type of design may be 'fixed' or 'random'. There should then be a second, 'random' factor nested within the standard factor.

An example

In an experiment on cholesterol level in blood of mice two levels of fat intake are fixed by the researchers and coded as level '0' and '1'. For each level of 'intake' there are three populations of mice in separate cages and from each of these cages three individual mice are selected at random for blood testing.

Chapter 7
The tests 1: tests
to look at
differences

The six populations of mice are assigned to the factor 'cage' which is a 'random' factor nested within 'intake' with the three cages in each group coded '1', '2' and '3' (clearly these labels have no real meaning and are only used for convenience). There are six cages in all but they must not be coded as 1 to 6. The data collected are as follows:

	Intake					
	0			1		
	Cage					
Mouse	1	2	3	1	2	3
1	55.6	62.5	58.9	85.6	68.5	65.6
2	62.5	68.3	54.2	98.2	69.3	71.0
3	68.2	58.2	63.5	75.1	88.2	78.3

Note: a common error made when analysing this type of data is that 'mouse' is treated as a factor. This seems sensible as the mice are labelled 1–3! This is inappropriate; as there is only one observation per mouse anova would have no variation to work with if it was used as a factor. Therefore, 'mouse' is actually the level of replication within the factor 'cage'. (However, if there were two observations per mouse then 'mouse' could be a further 'random' factor that would be nested within 'cage'.)

Using the example data the following ANOVA table should be used in a report:

Source of variation	Degrees of Freedom	Sum of Squares	Mean Square	F ratio	P value
Intake	1	1215.24	1215.24	18.90	0.001**
Cage within intake	4	377.42	94.35	1.47	0.272
Error	12	771.68	64.31		
Total	17				

SPSS. Nested analyses are achievable but slightly awkward in SPSS. Put the observations in a single column. Use two further columns for the factors with the levels coded as integers. From the 'Statistics' menu select 'ANOVA models' then 'General factorial...'. Put the observations into the 'Dependent variable:' box and the two factors in the 'Factor(s):' box. Use the 'Define Range...' button to input the highest and lowest group codes for each factor. (A detour

Chapter 7
The tests 1: tests
to look at
differences

via the 'Options…' button can be useful as means by factor levels can be requested allowing comparison of groups.)

Now the awkward bit… click on 'Paste'. This brings up the command line '!syntax' window of SPSS. For the example it contains the following:

```
MANOVA
   data  BY cage(1 3) intake(0 1)
   /NOPRINT PARAM(ESTIM)
   /METHOD=UNIQUE
   /ERROR WITHIN+RESIDUAL
   /DESIGN  .
```

Data output 7.54

To carry out a nested ANOVA the '/design' line must be altered to specifically request it. The nested factor must be specified as being within ('W' in SPSS) the main factor. The error term used to calculate the *F* ratio must also be given (the bottom line error in SPSS is also coded as 'W' for within-group). In the example the last line becomes '/DESIGN cage W intake vs W, intake vs W.' with a comma separating the two lines. To run this design from the '!Syntax' window click on the 'play' button (an icon of a small black triangle pointing to the right).

The following output appears:

```
* * * * * * A n a l y s i s   o f   V a r i a n c e -- design  1 * * * * * *

Tests of Significance for DATA using UNIQUE sums of squares
Source of Variation         SS      DF      MS        F   Sig of F

WITHIN CELLS             771.68     12    64.31
CAGE W INTAKE            377.42      4    94.35     1.47    .272
INTAKE                  1215.24      1  1215.24    18.90    .001

(Model)                 1592.66      5   318.53     4.95    .011
(Total)                 2364.34     17   139.08

R-Squared =          .674
Adjusted R-Squared =  .538
```

Data output 7.55

This output contains all the information required for the 'ideal' ANOVA table in a slightly different format. The error line is given first and called 'within cells' then come the two factors. The '(model)' line and below give summary information not usually required. In the example the *P* values (labelled 'Sig of F') show that there is no effect of 'cage' but there is a significant effect of 'intake' as P is very much less than 0.05. The degrees of freedom for nested factors are slightly odd as they have one less than the number of levels for each level of the main factor (in the example this is $(3-1) \times 2 = 4$).

A complication. If the nested factor has a *P* value less than 0.05 it is customary to use the mean square of this factor as the denominator to calculate the *F* ratio of the main factor rather than the error mean square. If your output shows that the 'Sig of F' for the nested factor is less than 0.05 you must return to the

[145]

Chapter 7
The tests 1: tests
to look at
differences

'!syntax' window and alter the '/design' line again as follows: '/DESIGN nestedf W mainf=1 vs W, mainf vs 1' replacing 'nestedf' and 'mainf' with the names of the relevant factors. This line now assigns the 'nestedf W mainf' to the number 1 and then tells the package to use that as a denominator for 'mainf' (hence 'vs 1' rather than 'vs W'). With more complicated designs selection of the correct denominators can become a rather long-winded process.

MINITAB

MINITAB. Put the observations in a single column. Use two further columns for the factors with the levels coded as integers. From the 'Stat' menu, select 'ANOVA' then 'General Linear Model…'. Move the observation column into the 'Responses:' box. Then put the main factor and the nested factor ('intake' and 'cage' in the example) into the 'Model:' box. To show that the nested factor is nested it must be followed by the name of the main factor in brackets (in the example the 'Model:' box reads: 'intake cage(intake)'). Click 'OK'. (A detour via the 'Options…' button can be useful as means by factor levels can be requested allowing comparison of groups.) (Or assuming data in c1, main factor in c2 and nested factor in c3 type 'GLM c1 = c2 c3(c2)' at the MTB> prompt in the session window.)

You get the following output from the example:

```
General Linear Model

Factor        Levels Values
intake            2    0     1
cage(intake)      3    1     2    3


Analysis of Variance for data

Source        DF    Seq SS    Adj SS    Adj MS      F      P
intake         1   1215.24   1215.24   1215.24  18.90  0.001
cage(intake)   4    377.42    377.42     94.35   1.47  0.272
Error         12    771.68    771.68     64.31
Total         17   2364.34
```

Data output 7.56

The output confirms the test used, the names of the factors, how many levels there are for each factor and the integer labels used. Check these are correct before looking at the ANOVA table. The table itself is very similar to the 'ideal' table given in the example. The degrees of freedom may seem a little odd in a nested analysis. The main factor will have one less degree of freedom than the number of factor levels as usual. The nested factor will have one less than the number of groups for each level of the main factor. In the example there are three cages, so there are two degrees of freedom for each of two levels of the main factor to give four degrees of freedom. The important column of the table is the final one with the P values (in the example there is no significant effect of 'cage' but a highly significant result for 'intake').

Excel

Excel. There is no easy way to carry out this type of nested analysis in Excel.

8: The tests 2: tests to look at relationships

Is there a correlation or association between two variables?

Observations assigned to categories

If observations are given a qualitative value then the data is said to be categorical (i.e. they have been assigned to categories). There are many occasions when it is not possible to express data in any other way (such as when scoring flower colours or species). There will be other circumstances when the categories are for convenience only, and are achieved by imposing a set of categories on a continuous scale. This can either be an arbitrary scale (e.g. the effect of an illness from 'well' through 'showing symptoms' to 'dead') or a scale that could be measured (e.g. dividing tree heights into 'short' and 'tall').

However, if the observations are assigned to categories in this way there are several tests that can be applied to determine whether the division of the observations into classes is independent or not. In other words 'are the observations for two categorical variables associated?'. The chi-square test of association, phi coefficient and Cramér coefficient are considered here.

Chi-square (χ^2) test of association

This is one of the most widely used statistical tests of all. It is beguilingly simple and has few underlying assumptions. In fact, the test is so simple to carry out it is often not properly supported in statistical packages, leaving the user to do a lot of the work. If observations can be assigned to one of two or more categories in two variables then chi-square is appropriate. The null hypothesis is that the categories in the two variables are independent (i.e. the category an observation is assigned to for one variable has no effect on the category it is assigned to for the second variable). For example if 'eye colour' and 'sex' are the two variables and individuals are assigned to either 'blue' or 'brown' and either 'male' or 'female' then the null hypothesis is that there is no association between 'sex' and 'eye colour'. As there are no assumptions made about the form of the data it is a nonparametric test although it is rarely described as such (probably because there is no parametric equivalent).

(*Note*: It is important to realise that if a continuous variable is forced into a small number of categories then information is lost when the test is calculated and another measure of association is probably more appropriate.)

The chi-square tests works by adding up the squared differences between the expected number of observations in a category combination and the actual observed number. The result is then looked up on a chi-square table with a

Chapter 8
The tests 2: tests
to look at
relationships

number of degrees of freedom equal to one less than the number of rows multiplied by one less than the number of columns.

Expected values are calculated very simply by putting the data into a table and totalling the observations in each row and column. The expected value for each category combination is the row total multiplied by the column total divided by the total number of observations.

There are complications. If there are expected values lower than one the test should not be used (categories should be combined to avoid this problem). No more than 20% of the expected values should be less than five.

Chi-squared tests should never be carried out on percentages or data transformed in any way. It must be carried out on frequencies (numbers of observations). When reporting the results in text it is usual to give the statistic, the degrees of freedom and some measure of the P value (e.g. $\chi^2 = 15.62$, d.f. $= 4$, $P < 0.01$).

An example

A group of students interested in riverine (riparian) invertebrates want to determine, in the shortest possible time, whether stream velocity is related to plant growth and substrate. A qualitative survey of fifty, randomly assigned stream sections in the study area was carried out. Stream velocity was scored as 'slow' or 'fast' (if more time were available then an accurate measure of the stream velocity could be made, although it might not be very useful as such a measure is very dependent on weather conditions). The stream bed was assigned to one of four categories: 'weed-choked'; 'some weeds'; 'shingle' and 'silt' (this is one of many possible classification systems that could be used).

The data was collected and compiled into a table as follows, numbers indicate the number of sites with each combination of categories:

Velocity category	Stream-bed category			
	Choked	Weeds	Shingle	Silt
Slow	10	8	2	7
Fast	2	6	10	5

The null hypothesis, H_0, is that the proportion of streams in each of the 'beds' categories is the same for the two velocity categories. The alternative hypothesis, H_1, is that the proportions vary between the two categories; i.e. there is some association with particular stream-bed categories and velocity categories appearing together more often than would be expected by chance and others appearing less often. The test will not determine which category combinations are more or less common than expected although this can be investigated by inspection of the table of raw data and expected values.

The data table needs to be processed to obtain the expected values. First the row and column totals are required.

Chapter 8
The tests 2: tests
to look at
relationships

	Choked	Weeds	Shingle	Silt	Totals
Slow	10	8	2	7	27
Fast	2	6	10	5	23
Totals	12	14	12	12	50

Then the expected value for each of the cells can be calculated. For 'Choked' and 'Slow' the expected value is the row total (27) multiplied by the column total (12) divided by the total number of observations (50). Of course, this sort of manipulation is very easy to do in a spreadsheet such as Excel.

The expected values are:

	Choked	Weeds	Shingle	Silt
Slow	6.48	7.56	6.48	6.48
Fast	5.52	6.44	5.52	5.52

In this example the value of chi-square is 11.036, there are 3 degrees of freedom and the *P* value is just over 0.01. This might be reported in the text of a results section as: a chi-square test showed there was an association between stream velocity category and stream bed category (χ^2 11.036, d.f. 3, $P<0.05$).

SPSS

SPSS. The chi-square test of association is not easy to achieve from a table of frequencies. However, if the data are arranged so that each individual is represented by a row with entries in two columns for the two factors then it is very easy to carry out.

Arrange the data in two columns with one column for each of the factors. Input the categories for each observation on a separate row with each category coded with integers starting at 1. The columns should be labelled appropriately.

The example data will have 50 rows and two columns set out using the following style:

	Velocity	Bed cat
1	1	4
2	2	2
3	2	3
:	:	:
50	1	2

Chapter 8
The tests 2: tests
to look at
relationships

The integer codes can be used as labels for the category names. This can be done in several ways in SPSS, although the easiest is to double-click on the variable name. This brings up a dialogue box. Click on 'Labels…'. In the 'Value labels' section type the code number in the 'Value:' box and the label you want to assign to it in the 'Value label:' box before clicking 'Add'. After assigning labels for all the categories click on 'Continue' then 'OK'. Doing this before running the chi-square test will make the output much easier to interpret.

To run the test go to the 'Statistics' menu, choose 'Summarize' and then 'Crosstabs…'. Move the column containing the 'row' information into the 'Row(s):' box (in the example this was 'velocity') and the other column into the 'Column(s):' box. Then click on the 'Statistics…' button. In the options box that appears select 'Chi-square' and click 'Continue'. Then click on the 'Cells…' button. In the options box select both 'Observed' and 'Expected' in the counts area. Click 'Continue' then 'OK' in the original box.

The following output will appear in the '!Output' window:

```
VELOCITY  by  BED_CAT

                      BED_CAT                        Page 1 of 1
              Count  |
              Exp Val|Choked   Weeds    Shingle  Silt
                     |                                    Row
                     |     1|      2|      3|      4| Total
VELOCITY      -------+-------+-------+-------+-------+
           1  |   10 |    8  |    2  |    7  |    27
    Slow      |   6.5|   7.6 |   6.5 |   6.5 |  54.0%
              +-------+-------+-------+-------+
           2  |    2 |    6  |   10  |    5  |    23
    Fast      |   5.5|   6.4 |   5.5 |   5.5 |  46.0%
              +-------+-------+-------+-------+
          Column      12     14      12      12      50
          Total     24.0%   28.0%   24.0%   24.0%  100.0%

     Chi-Square                    Value         DF         Significance
     ----------------------        ----------    ----       ------------

Pearson                           11.03635        3             .01153
Likelihood Ratio                  11.94537        3             .00757
Mantel-Haenszel test for           3.15996        1             .07547
     linear association

Minimum Expected Frequency -       5.520


Number of Missing Observations:   0
```

Data output 8.1

If the labelling of the categories had not been carried out there would only be numbers for the rows and columns. The table gives the number observed above the number expected for each category combination. Row and column totals are also given both as numbers of observations and as percentages. Then comes the results of the test itself. The important line is that labelled 'Pearson'. The number in the 'Value' column is the value of the X^2 approximation to χ^2 (11.036 in the example). Then the degrees of freedom are given (labelled 'DF') and finally the P value (labelled 'Significance'). Ignore the other two lines in this table. In the example the P value is less than 0.05 so the null hypothesis is rejected. Inspection of the table shows

Chapter 8
The tests 2: tests
to look at
relationships

that the biggest difference between observed and expected values are in the 'Shingle' column suggesting that there is an association of shingle stream beds with fast flowing water.

The minimum expected frequency (given as 5.520 in the example) is an important number to look at, as if it is less than 1 then the chi-square test should not be carried out. Caution should also be applied if more than 20% of the cells have expected frequencies less than 5. As shown below, SPSS gives a warning if there is a problem (although it still reports the results of the test and gives a *P* value).

```
Minimum Expected Frequency -    .471
Cells with Expected Frequency < 5 -    2 OF    10 ( 20.0%)
```

Data output 8.2

MINITAB. The chi-square test is easy to carry out from a table of counts of observations. First input the table exactly as it is laid out in the example. The columns can be labelled with the categories but there is no way to label the rows. Then from the 'Stat' menu, select 'Tables' then 'Chisquare test...'. In the dialogue box highlight the relevant columns and click on 'select'. This moves the columns into the 'columns containing the table:' box. Click 'OK'.

The example data without column labels gives the following output in the session window:

```
Chi-Square Test

Expected counts are printed below observed counts

            C1       C2       C3       C4     Total
    1       10        8        2        7        27
          6.48     7.56     6.48     6.48

    2        2        6       10        5        23
          5.52     6.44     5.52     5.52

Total       12       14       12       12        50

ChiSq =  1.912 +   0.026 +   3.097 +   0.042 +
         2.245 +   0.030 +   3.636 +   0.049 = 11.036
df = 3,  p = 0.012
```

Data output 8.3

The output confirms the test then gives a repeat of the table of frequencies but with the addition of row and column totals as well as the expected values. The section 'ChiSq' gives the value for: '(observed – expected)^2/expected' for each cell in the table and then gives the total of these values which is the statistic X^2 (estimating χ^2). The degrees of freedom are given and then the *P* value, labelled 'p'. In the example the *P* value is less than 0.05 so the null hypothesis is rejected. There is some association between stream-bed category and velocity category. Inspection of the 'ChiSq' values shows that the biggest contributors are in column 3 ('shingle') and show that a shingle bed is far

Chapter 8
The tests 2: tests
to look at
relationships

more common in a fast flowing stream and far less common in a slow flowing one than expected if there were no association between the factors.

Excel. The spreadsheet capabilities of Excel make the chi-square test quite easy to carry out.

1   Input the frequency data (number of observations in each category combinations) as a table with row and column labels exactly as in the example.

2   Add labels for an extra row and column of totals. In the relevant cells type '=SUM' and the cell numbers required (in the example the row totals column is F so the command in cell F2 is '=SUM(b2:f2)' and the first of the column totals is in B4 with the command '=SUM(b2:b3)'). Once the first cell has been calculated for row and column it can be pasted to calculate the other totals. An easy way of doing this is to highlight the cell then click and hold the small black square in the bottom right corner then drag across to highlight the cells to paste into. Make sure the total of row totals is calculated as well as this is the total number of observations.

3   A second table, the same size as the first should be labelled elsewhere on the same sheet. This is for the expected values. Once the row and column labels have been created the expected values for each cell can be calculated as '=row total * column total/number of observations' inputting the relevant cell numbers as required. In the example the top right cell expected value is calculated as '=b4*f2/f4'.

(It is useful to use the '$' as an anchor when calculating the expected values. Any cell code containing a '$' will *not* alter when the cell contents are pasted elsewhere on the spreadsheet. So when inputting the first cell calculation use '=b$4*$f2/f4' and then paste the cell across the whole table. The row containing column totals and the column containing the row totals as well as the cell containing the total number of observations are all fixed.)

This should be on the spreadsheet:

Observed	Choked	Weeds	Shingle	Silt	Totals
Slow	10	8	2	7	27
Fast	2	6	10	5	23
Totals	12	14	12	12	50

Expected	Choked	Weeds	Shingle	Silt
Slow	6.48	7.56	6.48	6.48
Fast	5.52	6.44	5.52	5.52

Data output 8.4

None of the expected values are below 5 so the test can proceed. If any cell had a value less than one or more than 20% have values less than 5 then categories should be combined.

Chapter 8
The tests 2: tests
to look at
relationships

4 Once the observed and expected tables are complete, select an empty cell anywhere on the spreadsheet where the two tables are still on the screen. Then go to the 'Function Wizard', select 'Statistical' and then 'CHITEST' from the list. Click 'Next'. Then put the cell codes for the observed data (not labels) into the 'actual_range' box. This can be done either by typing in the cell codes for top left and bottom right directly with a colon (:) between them or by selecting the 'actual_range' box and then clicking and dragging over the cells. Do the same for the 'expected_range' box selecting the expected values. Then click 'Finish'. The *P* value for the test appears. In the example the *P* value is '0.0115', which is lower than the critical '0.05' value so the null hypothesis is rejected. There is an association of 'stream-bed' category with 'velocity category'. Comparison of the observed and expected values indicate that 'Slow' and 'Choked' are more common than would be expected by chance as are 'Fast' and 'Shingle'.

5 It is usual to present the results of a chi-square test giving the degrees of freedom and the value of χ^2. Degrees of freedom for the test is one less than the number of rows multiplied by one less than the number of columns (in the example $(4-1)*(2-1)=3$). Excel provides a simple method for determining the χ^2 value from a known *P*-value and degrees of freedom. Select an empty cell. Click on the 'Function Wizard', select 'Statistical' then 'CHINV'. Click on 'Next'. Input the *P*-value into the 'probability' box and input the degrees of freedom, then click 'Finish'. The χ^2 value appears in the cell.

Cramér coefficient of association

A test carried out on tables of frequencies in conjunction with a chi-square test that provides additional information about the strength of the association. The statistic X^2 is used to determine significance, while the Cramér coefficient (*C*) is a measure from 0 (no association) to 1 (perfect association) that is independent of the sample size. If this statistic is available as an option on the package you are using then it will allow direct comparison of the degree of association between tables.

Phi coefficient of association

A special case of the Cramér coefficient for 2×2 tables (i.e. there are only two categories for each of the two variables). It is used in combination with the chi-square test as it generates a value (r_ϕ) that ranges from 0 to 1, indicating a range from no association through to perfect association. It can be calculated directly after a chi-square test has been carried out as it is equal to the square root of the value of X^2 after it has been divided by the number of observations.

Observations assigned a value

If each individual is assigned a meaningful numerical value in two variables

Chapter 8
The tests 2: tests
to look at
relationships

there are several tests that may be applied to determine whether the two sets of observations are associated or correlated, the strength of the correlation and whether it is significant or not. Four tests are considered here, the Pearson's product–moment correlation, the Spearman's rank-order correlation, the Kendall's rank-order correlation and regression. The most appropriate test is determined by the distribution and quality of the data.

'Standard' correlation (Pearson's product–moment correlation)

When a correlation is mentioned it is almost invariably the Pearson product–moment correlation that is in mind. The statistic (r to estimate the true correlation, ρ) produced by the test ranges from –1, through 0, to 1 and describes a range of associations between two variables from 'perfect negative correlation', through 'no correlation' to 'perfect positive correlation'. This test is very widely applied, perhaps too widely as it has some rather severe assumptions about the distribution of the two variables being investigated. Both variables must be measured on a continuous scale and both must be normally distributed. If these assumptions do not apply, the Spearman's rank-order correlation should be used instead.

When quoting the results of this test in a report it is usual to use a scattergraph (without trend line) and the form: 'Pearson product–moment correlation indicates a significant positive association between x and y ($r=0.51$, d.f. $=22$, $P<0.05$)'.

Two words of caution:

1 It is quite rare to find two variables that are normally distributed and therefore suitable for Pearson's correlation. Consider the alternatives.

2 The statistical significance of correlation is not a good guide to the *real* significance of the correlation. With large sample sizes the value of r required to achieve statistical significance (i.e. to show that there is some relationship between the two variables) is rather low. It is perhaps better to use the value of r^2 as an indicator of the real significance as this value shows the amount of variation in one variable explained by the other.

An example

A marine biologist working on rockhopper penguins has measured the sizes of birds forming pairs. The measure used is the length of a bone in the leg which is known, from previous studies, to be a good indication of size. It is measured to the nearest 0.1 mm. The null hypothesis is that male size is not correlated with female size. Unfortunately data could only be collected from six pairs.

Chapter 8
*The tests 2: tests
to look at
relationships*

Pair	Female	Male
1	17.1	16.5
2	18.5	17.4
3	19.7	17.3
4	16.2	16.8
5	21.3	19.5
6	19.6	18.3

It is assumed that both variables are normally distributed. This data set is a little small to prove this, although a larger sample of the population might be (see 'Do frequency distributions differ?' in the previous chapter, page 61). In this case the null hypothesis is rejected as there is a significant positive correlation between male and female size indicating that there is positive assortative mating in this species. The value of r is 0.88 and r^2 is 0.77 indicating that 77% of the variation in the size of one sex is explained by the size of the other.

SPSS

SPSS. This is one of the easiest tests to carry out in SPSS. Input the data in two columns and add appropriate column labels. The cases (pairs in the example) do not require a separate column. From the 'Statistics' menu, select 'Correlate' then 'Bivariate...'. In the dialogue box move the names of the two variables into the 'Variables:' box and make sure that the 'Pearson' option is checked. Click 'OK'.

The following output appears:

```
- -  Correlation Coefficients   - -

                    FEMALE      MALE

FEMALE            1.0000       .8813
                 (    6)      (    6)
                 P= .         P= .020

MALE               .8813      1.0000
                 (    6)      (    6)
                 P= .020      P= .

(Coefficient / (Cases) / 2-tailed Significance)

" . " is printed if a coefficient cannot be computed
```

Data output 8.5

The output gives a matrix of correlation coefficients, degrees of freedom and P values although only one set is actually useful. The correlations of 'x' with 'x' and 'y' with 'y' (e.g. female with female in the example) unsurprisingly report a perfect correlation of '1.0000'. The important section is the correlation of 'x' with 'y' (female with male) in the example. The

Chapter 8
The tests 2: tests
to look at
relationships

statistic is given first ($r = 0.8813$ in the example), then the numbers of pairs of observations in brackets and finally a *P* value (labelled P= and 0.020 in the example). As *r* is positive it shows that there is positive size assortment of individuals in pairs in the example (i.e. large females tend to be paired with large males) and the association is strong enough to reject the null hypothesis and accept the alternative hypothesis.

MINITAB

MINITAB. Input the data in two columns using one row for each pair of observations. Label the columns appropriately. From the 'Stat' menu select 'Basic statistics' then 'Correlation…'. Highlight the names of the two variables then press the select button (Fig. 8.1). Click 'OK'. (Or type 'corr c1 c2' at the MTB> prompt in the session window.)

The following output appears:

```
Correlations (Pearson)

Correlation of female and male = 0.881
```

Data output 8.6

The test reports the value of *r* and nothing else. In the example this indicates that there is a strong positive association between the two variables — indicating that there is positive assortative mating. If you want to

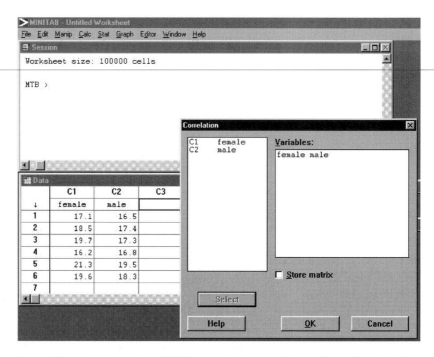

Fig. 8.1 Bivariate correlation in MINITAB. Each pair of variables should be highlighted and selected to move them into the 'Variables:' box.

Chapter 8
The tests 2: tests
to look at
relationships

know the P value associated with this value of r or determine r^2, this can be done using regression (regress one variable on the other and inspect the r^2 value and P value reported in the ANOVA table section of the output; see the regression section later in this chapter for further details).

Excel. There are two methods, the first requires the 'Analysis ToolPak' to be installed, the second does not.

Method 1. Arrange the data in two columns with appropriate column labels. From the 'Tools' menu select 'Data analysis…' then 'correlation'. Click 'OK'. The 'input' box should contain the cell code for the top left and bottom right of the area containing the data separated by a colon (':'). This can be done most easily by selecting the 'input' box then clicking on the top left cell and dragging to the bottom right. If the variables are included in the cells defined the 'Labels in first row' box should be checked. The output can either appear on a separate sheet or elsewhere on the same sheet. If on the same sheet the 'Output range' option needs to be selected and the cell indicating the point at the top left of the output put into the box. Click 'OK'.

This table appears:

	Female	Male
Female	1	
Male	0.88132	1

Data output 8.7

Method 2. Arrange the data in two columns (or two rows). Select an empty cell then click on the Function wizard (f_x) and select 'Correl'. Click 'Next'. Input the range of cells containing the data for the first variable into the 'array1' box (this can be done by either typing in the first and last cell with ':' between them or by clicking on the first cell and dragging the pointer to the last cell). Put the range for the second variable in the 'array2' box and click 'Finish'. The r value appears in the cell.

Either method will produce the same number although the output from method 1 is more extensive. The value of the statistic r is given. It is 0.88 using the example data indicating strong positive association of the two sets of observations. The r^2 value can be determined by selecting an empty cell and typing '=0.88^2' (the '^' symbol is often used to mean 'to the power of'). If the P value associated with this value of r is required this can be determined using regression (regress one variable on the other and inspect the P value reported in the ANOVA table section of the output; see the regression section later in this chapter for further details).

Chapter 8
The tests 2: tests
to look at
relationships

Spearman's rank-order correlation

This is one of two commonly used nonparametric equivalents of the Pearson's product–moment correlation. The statistic it gives is called r_s and ranges from −1 through 0 to 1, indicating 'perfect negative correlation', 'no correlation' and 'perfect positive correlation'. Although this is the same as the Pearson r it is not advisable to compare the results from this test directly with values of r.

As long as there are two observations for each individual and that the observations are measured on a scale that can be put into a meaningful rank order this test is appropriate. Spearman's correlation is much more conservative than Pearson's.

An example

The same penguin example will be used as for the Pearson product–moment correlation used earlier. It is quite possible that a researcher will use the Spearman's rank order correlation rather than the Pearson if there is a doubt that the data are appropriate for the Pearson test.

SPSS

SPSS. Arrange the data in two columns with each row representing an individual (in the example the 'individuals' being considered are actually pairs of birds). Label the columns appropriately. From the 'Statistics' menu select 'Correlate' then 'Bivariate...'. Highlight the names of the two variables and move them into the 'Variables:' box. Make sure that 'Spearman' is checked (Pearson is checked by default). Click 'OK'.

The following output appears:

```
- - -  S P E A R M A N   C O R R E L A T I O N   C O E F F I C I E N T S  - - -

MALE              .7714
             N(     6)
             Sig .072

             FEMALE

(Coefficient / (Cases) / 2-tailed Significance)

" . " is printed if a coefficient cannot be computed
```

Data output 8.8

This confirms the test used and then gives the value of r_s, the number of pairs of observations and the *P*-value associated with the r_s value (labelled 'Sig'). In the example, although the statistic is clearly showing a large positive association, it is not large enough to be deemed significant because the sample size is rather small. However, the *P*-value is sufficiently close to 0.05 to warrant further investigation.

Chapter 8
The tests 2: tests
to look at
relationships

MINITAB

MINITAB. This test cannot be carried out directly. The data first have to be put into rank order and then a Pearson correlation carried out on the ranked data.

1 Input the data in two columns. Label the columns appropriately. If there are any individuals where one or both of the observations are missing these should be omitted from the data set. From the 'Manip' menu select 'Rank…'. Put the name of the first variable into the 'Rank data in:' box and put a new name in the 'Store ranks in:' box. Click 'OK'. A new column will appear with the original data replaced with integers starting at '1' for the smallest value. Repeat for the second variable.

2 Carry out a normal correlation on the *ranked* data. From the 'Stat' menu select 'Basic statistics' then 'Correlation..'. Highlight the two *ranked* variables in the list on the left and click on select to move them into the 'Variables:' box. Click 'OK'.

This output appears:

```
Correlations (Pearson)

Correlation of Rfemale and Rmale = 0.771
```

Data output 8.9

The test indicates that a Pearson correlation has been carried out (the package does not 'know' that ranked data are being used). The correlation reported is the value of r_s. In this case it shows that there is a strong positive association between male and female size. The significance of the relationship is not given.

Excel

Excel. There is no direct way of carrying out this test even with the Analysis ToolPak installed. However, it is possible to carry out the test by first ranking the data and then performing a normal correlation.

Input the data in two columns with a pair of observations on each row. Do not include any rows where one or both of the observations are missing. Use the cells at the top of the columns for appropriate labels. To rank the data, first select a cell in an empty column in the same row as the first row of data. To use the 'rank' operation it is probably easiest to simply type '=RANK(A2,A\$2:A\$7)' rather than use the Function Wizard (f_x). This command asks that the cell is given the rank of cell A2 in the range of cells A2 to A7 (the six cells containing the female sizes in the example). The inclusion of the '\$' symbol makes it very easy to use the copy and paste commands to rank the whole data set as it anchors the row numbers. First select the cell, then click on the small black square in the bottom right and drag to the cell in the next column to the right level with the *last* observation in the data set. This will place the appropriate 'RANK' syntax in each cell without having to type in anything else (in the example it left '=RANK(B7,B\$2:B\$7)' in the bottom right cell). Label the columns containing the ranks appropriately.

Chapter 8
The tests 2: tests
to look at
relationships

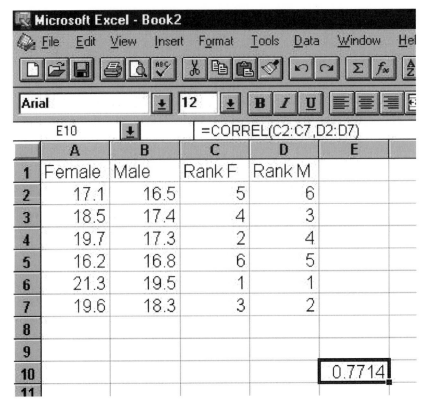

Fig. 8.2 Spearman's rank correlation in Excel. The data have to be ranked first and then a normal (Pearson's) correlation is carried out on the ranked data.

Carry out a Pearson correlation on the *ranked* data. Select an empty cell. Click on the Function Wizard (f_x) and select 'CORREL' from the 'statistics' section of the list. Click 'Next'. Input the range of cells containing the data for the first variable into the 'array1' box (this can be done by clicking on the first cell and dragging the pointer to the last cell). Put the range for the second variable in the 'array2' box and click 'Finish' (Fig. 8.2). The r_s value appears in the cell.

In the example the value of r_s is 0.7714 indicating a strong positive association of ranked female and ranked male size. The significance is not given.

Kendall's rank-order correlation

A second, and slightly less widely used, nonparametric correlation. It is sometimes called Kendall's Tau and the statistic produced is usually denoted as *T*. As with the Spearman's correlation the test can be carried out on any data set where there are two observations for each individual and the data can be put into a meaningful rank order. Like other measures of association, *T* ranges from −1 through 0 to 1, indicating 'perfect negative correlation', 'no correlation' and 'perfect positive correlation'. Although this is the same as

Chapter 8
The tests 2: tests
to look at
relationships

the Pearson *r* and the Spearman r_s it is not advisable to compare the results from this test directly with either.

The only slight advantage of Kendall's correlation over Spearman's is that *T* can be used in partial correlation whereas r_s cannot.

> It is advisable to use caution when interpreting the significance of a value of *T* as when sample sizes are very large there is likely to be a *P* value less than 0.05 even though the association is only slight.

Example

The same example as used for the Pearson and Spearman tests is used.

SPSS

SPSS. As for Spearman's correlation, arrange the data in two columns with a row for each 'individual'. Label the columns appropriately. From the 'Statistics' menu select 'Correlate' then 'Bivariate…'. Select the two variables and move them into the 'Variables:' box. Select the 'Kendall's tau-b' option from the 'Correlation coefficients:' list. Click 'OK'.

The following output appears:

```
- - - K E N D A L L   C O R R E L A T I O N   C O E F F I C I E N T S   - - -

MALE            .6000
            N(    6)
            Sig .091

            FEMALE

(Coefficient / (Cases) / 2-tailed Significance)

" . " is printed if a coefficient cannot be computed
```

Data output 8.10

This confirms the test used and then gives the value of *T*, the number of pairs of observations and the *P* value associated with the *T* value (labelled 'Sig'). In the example although the statistic is clearly showing a large positive association it is not large enough to be deemed significant because the sample size is rather small. However, the *P* value is sufficiently close to 0.05 to warrant further investigation.

Note that the value of *T* is rather less than the value of r_s in the Spearman's test using the same data. This highlights the fact that the two statistics should not be compared directly.

MINITAB

MINITAB. There is no direct way of carrying out this test.

Excel

Excel. There is no direct way of carrying out this test.

[161]

Chapter 8
The tests 2: tests
to look at
relationships

Regression

The use of regression usually implies that a prediction of one value is being attempted from another: e.g. cause and effect. Regression analysis is considered in the next section. However, as most regression outputs give a P value and an r^2 value the association with the Pearson's correlation is very strong. The P value given in a standard linear regression is the probability that the best-fit slope of the relationship between two variables is actually zero. In a comparison with the Pearson statistic this translates to the probability that there is no relationship (i.e. $r = 0$).

The advantage of using regression rather than Pearson's correlation is that the assumption that both variables are distributed normally is lifted. The assumptions are different, although slightly less restrictive. For example, regression assumes that the x ('cause') values should be measured without error; that the variation in the y ('effect') is the same for any value of x and for linear regression that the relationship between two variables can be described by a straight line. Of these the assumption that variance in y is the same for all values of x is probably the least likely to be true, it is usual for variance in y to increase as the x ('cause') variable increases.

If you decide to use regression to determine the association between two variables please use great caution because the implication is that one of the variables in some way depends on the other. Also, one of the underlying assumptions of regression is that the values of the 'cause' variable are in some way set or chosen by the investigator; clearly this is not the case if the observations are taken at random.

Example

Again we consider the penguin pairs used as the example throughout this section. The researchers were testing the null hypothesis that male and female birds were forming pairs independent of their size. The alternative hypothesis was that there was an association (either positive or negative) of male and female sizes in pairs. Framed in this way the hypothesis is not suitable for regression, but if the penguins form pairs by a choice of one sex for another then it might become more like a regression problem. If females actively chose males then the null hypothesis can be framed in regression terms as 'male size does not depend on female size'. However, even if male size can be said to *depend* on female size there is still the problem that the female sizes were not set or chosen by the investigator.

Is there a 'cause and effect' relationship between two variables?

Questions. *There are many circumstances where it is clear that one set of observations in some way depends on another. In this section there will be two observations for each 'individual' with one observation being the 'cause', x, 'predictor' or 'independent' variable that is set or chosen by the experimenter*

Chapter 8
The tests 2: tests
to look at
relationships

and the other being the 'effect', y or 'dependent' variable which is never set by the experimenter. There are a variety of methods that can be applied to determine the form and strength of the relationship between the 'cause' and 'effect' that make different assumptions about the variables and the form of the relationship between them. Five tests are considered here: linear regression; Kendall's robust line-fit method; logistic regression; model II regression and polynomial regression.

'Standard' linear regression (a.k.a. model I linear regression)

This is a very widely used statistic in biology. It is also, possibly, the most abused statistic in biology. Linear regression is an extremely powerful and useful technique that determines the form and strength of a relationship between two variables. It is used if the intent is to be able to predict a value for y ('effect' or 'dependent') from a given value of x ('cause', 'predictor' or 'independent'). There are several components to any output. The 'slope' is the slope of a straight line of best fit drawn through the set of points with coordinates defined by the two variables. Slope can be positive or negative indicating an increase or decrease of y with increasing values of x. The slope can theoretically take any value. The second component of the output is the 'intercept' or 'constant'. This is the predicted value of y when x is equal to zero. The slope is often called b and the intercept a or c or 'constant'. Both slope and intercept are usually quoted with some measure of their variability (e.g. a 95% confidence interval or standard deviation). There is often a significance test result given with regression output. This is a test of whether the slope is zero or not. If the P value is less than 0.05 this should be interpreted as an indication that the slope is significantly different from zero, showing that there is a relationship between the x and y variables.

Linear regression makes many assumptions about the data sets. Important assumptions include: that the values of x are measured without error; that the values of x are chosen or set by the experimenter; that the relationship between x and y is best fitted by a straight line ($y = a + bx$) and that the variation in y is the same for all values of x.

> *Tip*: if you are unsure which of the two variables is the x ('cause') variable then linear regression is almost certainly not appropriate.

Prediction

Once a best-fit line has been determined then a value for the 'effect' can be predicted for any value of the 'cause'. In practice it is unwise to use values of the 'cause' variable that are beyond the range of the data used to fit the line as the shape of the relationship is unlikely to be the same across all values of the 'cause'. Although it may appear tempting you should *never* use an observation of the 'effect' to make a prediction of the 'cause' by simple rearrangement of the algebra of the best fit line.

[163]

Chapter 8
The tests 2: tests
to look at
relationships

Comparison of regression and correlation

The assumptions of linear regression are very different to those of correlation. 'Standard' correlation assumes that both x and y are normally distributed. For this reason it is tempting to use regression in lieu of correlation when variables are not normally distributed. Do not do this—try an alternative correlation instead because regression makes different assumptions that are unlikely to be true if correlation was the preferred statistical technique.

Residuals

The variation in y not accounted for by the best fit line relationship between x and y is called the residual variation. For each observation of x there is a predicted value on the line of y. Because y varies the point is not going to lie exactly on the line and the *vertical* distance from the point to the line is the *residual* for that point. It is often helpful to examine the residuals by plotting them against x. This is offered as an option in most statistical packages. If the relationship of x and y is really a straight line or there is no relationship at all then the residuals will be scattered without pattern for all values of x. However, if the relationship is really a curve then the residuals of a best fit straight line will show this to be the case—most of the residuals will be negative (or positive) at the ends of the line and positive (or negative) in the middle. Also, one of the assumptions of regression is that there is the same variation in y for all values of x. If the residuals are all small at one end of the range of x values and large at the other this would indicate that this assumption had been violated.

Confidence intervals

The confidence intervals attached to slope and intercept allow a range of possible lines to be drawn with boundaries enclosing the range within which 95% (or 90% if preferred) of the lines of best fit will appear. As this range of possible lines is in some way anchored to the y-axis at the point where x is zero and comprises a range of lines of different slope the 95% confidence limits of the lines are not straight lines. They are always curved lines which are closer to the best fit line in the middle of the range of values of x and further away from it at the extremes.

Prediction intervals

Prediction interval is the range within which 95% of the values of y are predicted to occur for any given value of x. The 95% prediction intervals (PI) will always be further away from the best-fit line than the 95% confidence interval for the line. They do not run parallel to the best-fit line but, like the confidence interval for the line, are narrower for mid values of x.

An example

A team of researchers is investigating the uptake of an experimental drug through the stomach. They suspect that the acidity of the stomach will affect uptake and propose an experiment to determine the uptake across a range of acidities. Using preparations of sheep stomach and a fixed concentration of the drug in solution of a range of pH is prepared and the passage of the drug through the stomach monitored. This is suitable for regression as the acidities ('cause') are set by the experimenters and we can assume that they are measured without error.

	pH					
	0.6	0.8	1.0	1.2	1.4	1.6
Uptakes	11.32 11.31 11.22 11.23	11.29 11.22 11.18 11.21	11.37 11.40 11.38 11.37	11.32 11.31 11.35 11.32	11.32 11.36 11.40 11.35	11.49 11.52 11.38 11.49

It is usually best to plot the data as a scatter graph at this point to get a feel for the form of the relationship. This has been done in Excel (Fig. 8.3).

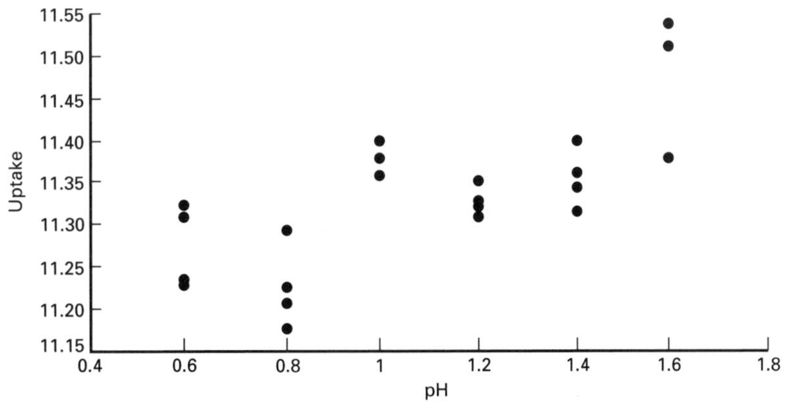

Fig. 8.3 Visualization of data suitable for regression using Excel. There were six pH levels set by the experimenter and four observations of uptake made at each level. This sort of plot gives a feel for the variation and trend in the data before any regression analysis is undertaken. It is right to plot the data like this before a trend line is added as the line will draw the eye.

SPSS. There are many options that can be applied to enhance the output of the standard linear regression in SPSS. First, the very minimum output will be considered and then some of the options that might be useful to help interpret the form and meaning of the relationship between the two variables.

Chapter 8
The tests 2: tests
to look at
relationships

Input the data in two columns and label them appropriately. There should be one row for each observation (two numbers). From the 'Statistics' menu select 'Regression' and 'Linear...'. Move the column containing the y or 'effect' variable into the 'Dependent:' box. Move the x or 'cause' variable (the one set or chosen) into the 'Independent(s):' box. Ignore all the routes to possible options at the moment, just click 'OK'.

Using the example data the following output appears:

```
* * * *   M U L T I P L E   R E G R E S S I O N   * * * *

Listwise Deletion of Missing Data

Equation Number 1    Dependent Variable..   UPTAKE

Block Number  1.  Method:  Enter     PH


Variable(s) Entered on Step Number
   1..    PH


Multiple R            .75769
R Square              .57409
Adjusted R Square     .55473
Standard Error        .05893

Analysis of Variance
                     DF      Sum of Squares      Mean Square
Regression            1            .10299            .10299
Residual             22            .07641            .00347

F =      29.65382      Signif F =  .0000


----------------- Variables in the Equation -----------------

Variable           B         SE B       Beta        T   Sig T

PH            .191786     .035219    .757686    5.446   .0000
(Constant)  11.126952     .040566             274.296   .0000


End Block Number   1   All requested variables entered.
```

Data output 8.11

Even this minimum level of output is rather bewildering! First the output confirms the test used (SPSS uses multiple regression which can cope with any number of 'cause' variables even when there is only one) and then explains how missing data is treated. Then comes the name of the 'dependent' or 'effect' variable followed by several lines giving the method that the 'cause' variable is entered into the analysis.

Then come four lines giving the strength of the relationship between y and x. This is similar to the r of correlation. The r^2 can be interpreted as the proportion of the variation in y explained by the best-fit line. In the example over half of the variation is explained by the line.

Next, is an ANOVA table comparing the best fit line against a null hypothesis of no relationship (e.g. slope is zero). In this table the F ratio and its associated P value (labelled 'Signif F' here) is given below the table. In the example the P value is given as '.0000' meaning '$P < 0.0001$' which is highly significant. There is very little doubt that a significant amount of the variation in y is explained by x.

Chapter 8
The tests 2: tests
to look at
relationships

Finally, comes the estimates for the two 'variables in the equation' (i.e. the slope and the intercept). The slope is given first on a row labelled 'PH' in the example. The estimated value for the slope is given, and is followed by an estimate of the standard error associated with the slope and then a *t*-test of the null hypothesis that the slope is equal to zero (this is essentially a repeat of the result of the ANOVA table). The second row refers to the intercept (labelled '(constant)') here giving a value with standard error and then a *t*-test of the null hypothesis that the intercept is zero.

In this example the slope of the relationship between 'pH' and 'uptake' is 0.1918 and the intercept (value of 'uptake' when 'pH' is zero) is 11.13. Both these values are significantly different from zero. The best fit line of 'uptake $= 0.1918 \times$ pH $+ 11.13$' explains 57% of the variation in the observations of 'uptake'. The best fit line can be, theoretically, applied to any value of x ('pH') to make a prediction of y ('uptake').

MINITAB

MINITAB. There are many optional extras that can be added to the basic regression analysis. Some of these extras are valuable for exploring the relationship between the x and y variables. I will consider the basic output first. Note that MINITAB calls the y or 'effect' variable the 'response' and the x or 'cause' variable the 'predictor'.

Put the data in two columns. One for the x values and one for the y. Label the columns appropriately. From the 'Stat' menu select 'Regression' then 'Regression...'. In the dialogue box that appears move the column containing the y values into the 'Response:' box and the column of x values into the 'Predictor(s):' box. Ignore the large array of additional options for the moment. Click 'OK'. (Or, assuming the y values are in c1 and the x values in c2 type 'Regress c1 1 c2' at the MTB> prompt in the session window.)

This output appears:

```
Regression Analysis

The regression equation is
Uptake = 11.1 + 0.192 pH

Predictor        Coef       Stdev      t-ratio        p
Constant       11.1270      0.0406      274.30      0.000
pH             0.19179      0.03522       5.45      0.000

s = 0.05893     R-sq = 57.4%     R-sq(adj) = 55.5%

Analysis of Variance

SOURCE         DF         SS         MS         F         p
Regression      1      0.10299    0.10299     29.65     0.000
Error          22      0.07641    0.00347
Total          23      0.17940
```

Data output 8.12

The output first confirms the test and then gives the best-fit line of the two variables. This describes both the intercept (11.1 in the example) and the slope (0.192 in the example). Then comes a section repeating the two parameters

Chapter 8
The tests 2: tests
to look at
relationships

where the intercept is labelled as 'Constant'. A standard deviation is given for each along with a *t*-test with the null hypothesis that the parameters are zero and the *P*-value (labelled 'p') associated with the *t*-test. In the example both slope and intercept are highly significantly different from zero. Next comes a line that reports the proportion of the variance in the *y* values that is explained by the best-fit line. In this case the r^2 value (labelled 'R-sq') is 57.4%.

Finally, comes an analysis of variance table that compares the variation explained by the best-fit line with the residual variation. The null hypothesis is that none of the variation in *y* is explained by the regression line. In this case the relationship is highly significantly different from zero and *P* is given as '0.000' although this should be reported as $P < 0.001$.

Excel

Excel. Regression is only possible if the 'Analysis ToolPak' has been installed. Put the data into two columns. One for the 'effect' and one for the 'cause'. Each row should represent a single observation (i.e. two values). Use the cells above the data to label the columns appropriately. From the 'Tools' menu select 'Data analysis…' then select 'Regression' from the long list and click 'OK'.

The 'Input y Range:' box should contain the code for the first and last cells with the list of *y* values (including the label). This can be most easily done if you click in the box to move the cursor there then click on the top of the list of data and drag to the bottom of the list. Repeat this for the 'Input x Range:' box with the *x* values (these are the ones set or chosen by the experimenter). If the labels have been included then make sure that the 'Labels' option is checked.

To define where the output is to appear choose either 'Output range:' and point at the top left corner of where the output will appear or leave the selection as 'New worksheet ply:' which will put the output in a fresh worksheet. Click 'OK'.

The example data produces this output:

```
SUMMARY OUTPUT

 Regression Statistics
Multiple R      0.755454
R Square         0.57071
Adjusted R      0.551197
Square
Standard        0.058694
Error
Observatio            24
ns

ANOVA
                 df        SS        MS        F      Significance F
Regression        1  0.100759  0.100759  29.24741    1.97E-05
Residual         22  0.075791  0.003445
Total            23   0.17655

             Coefficient Standard  t Stat   P-value    Lower      Upper     Lower      Upper
                 s       Error                95%        95%      95.000%   95.000%
Intercept    11.12937  0.040402  275.468  2.04E-40  11.04558  11.21316  11.04558  11.21316
pH            0.189698  0.035077  5.408087  1.97E-05  0.116953  0.262442  0.116953  0.262442
```

Data output 8.13

Chapter 8
The tests 2: tests
to look at
relationships

This output is rather confusing. First the output confirms that Regression statistics are being reported. Then come four lines giving the strength of the relationship between *y* and *x*. This is effectively the *r* of the Pearson's correlation. The 'R Square' (usually given as r^2) can be interpreted as the proportion of the variation in *y* explained by best-fit line. In the example 57% of the variation in *y* is explained by the line.

Next is an ANOVA table comparing the best fit line against a null hypothesis of no relationship (e.g. slope is zero). In the example the *P* value (labelled as 'Significance F') is given as 1.97E-05 meaning 0.0000197 or $P < 0.0001$ which is highly significant. There is very little doubt that a significant amount of the variation in *y* is explained by *x*.

Finally come the estimates for the two parameters in a separate table (i.e. the slope and the intercept). The first row refers to the intercept giving a value with standard error and then a *t*-test of the null hypothesis that the intercept is zero with associated *P* value (ludicrously small in the example) and then estimates of the 95% confidence intervals for the intercept (in the example the intercept is 95% likely to lie between 11.04 and 11.21). The slope is given on a row labelled 'pH' in the example. The estimated value for the slope is given, and is followed by an estimate of the standard error associated with the slope and then a *t*-test of the null hypothesis that the slope is equal to zero with *P* value (this is essentially a repeat of the result of the ANOVA table) followed by 95% confidence intervals for the slope.

In this example the slope of the relationship between 'pH' and 'uptake' is 0.1918 and the intercept (value of 'uptake' when 'pH' is zero) is 11.13. Both these values are significantly different from zero. The results can be interpreted as a best fit line of 'uptake $= 0.1918 \times pH + 11.13$' explaining 57% of the variation in the observations of 'uptake'. This best fit line could be, theoretically, applied to any value of 'pH' to predict 'uptake'.

Kendall's robust line-fit method

This is a simple nonparametric test that can be used instead of normal regression. It is unlikely to be supported by a statistical package, although it is likely to be possible to calculate using simple spreadsheet manipulations. The idea is to calculate the slope of the line between every possible pair of *x*, *y* points, and then to use the median value of the list of slopes as the best-fit slope.

This test makes very few assumptions about the data other than it is measured on a meaningful scale.

Logistic regression

A special form of regression analysis that is used when the 'dependent' variable can only be classified into groups (many packages limit this to two groups) is called logistic regression. It is a regression analysis that uses the proportions of the two possibilities of the 'dependent' to calculate the relationship with the 'independent' or 'cause' variable.

Chapter 8
The tests 2: tests
to look at
relationships

Model II regression

The model II regression is actually a group of analyses that make far fewer assumptions about the data than standard, model I regression. The main culprits are the assumptions in standard regression that the x values are measured without error and that the variation in y is the same for any level of x. If these assumptions do not hold then model II regression is appropriate. Unfortunately the statistical manipulations required are still being developed and there are several techniques that are not wholly satisfactory. Moreover, they are unlikely to be supported by computer packages.

Polynomial and quadratic regression

One of the assumptions of standard regression is that the form of the relationship between x and y is a straight line. If this is not true then the first option is to transform either x or y to make the data better fit a straight line. If this does not help then the assumption can be discarded and polynomial regression (of which quadratic regression is a special case) can be employed. The only difference between polynomial and linear regression is that the best-fit line is not straight. The advantage is that a curved line is almost always a better fit allowing a better prediction of y for each value of x. The disadvantage is that extra parameters have to be included. There is no longer a single 'slope' value but rather two or more factors that have to be applied to x, x^2, x^3 etc. It is usually either simply a scatter plot or inspection of the residuals from a linear regression which indicate that the straight line is not the best fit and that some sort of curve should be tried.

Tests for more than two variables

Tests of association

Questions. *Most tests of association assume that there are only two variables being considered. However, it is often the case that for each 'individual' there are three or more observations. If all that is required is to investigate associations in pairs to examine the strength of associations when other variables are accounted for, then this section considers briefly correlation, partial correlation and its nonparametric equivalent Kendall's partial rank-order correlation. Once there are several variables for each individual then multivariate analyses considered in the next chapter will often be more appropriate.*

Correlation

The simplest method used to analyse data sets with more than two observations for each 'individual' or 'site' is to consider all the possible two-way comparisons that could be made. Statistical packages allow this and will

Chapter 8
The tests 2: tests
to look at
relationships

happily produce a large matrix of correlation coefficients that can be trawled to find the largest positive and negative numbers for further investigation.

> *Note*: if the package reports a significance value for each of the correlation coefficients then they should be treated with some caution. Remember that the *P* value is just the probability that the null hypothesis is true and we usually reject the null hypothesis when $P < 0.05$. This implies that twenty correlations will produce, on average, one *P* value that is less than 0.05 even when there is no association at all. Therefore if there are 10 variables for each 'individual' giving 45 possible pairwise correlation coefficients the chances are very high that one or more will report *P*-values less than 0.05.

See the section on correlation for the mechanics of using the packages (pages 149–153).

Partial correlation

A partial correlation coefficient gives a measure of the relationship between two variables when one or more other variables have been held constant. A common use of this technique is in morphological analysis when several variables all relate in some way to size and therefore the bivariate correlation matrix technique employed in the previous section merely confirms that all measures are strongly correlated with each other. The partial correlation of two variables when 'size' is held constant will reveal whether they are related in any other way.

Kendall's partial rank-order correlation

The Kendall's rank correlation coefficient is the nonparametric equivalent of Pearson's product–moment correlation coefficient which can be used in partial correlation analysis. If variables are known to be distributed in a non-normal fashion then this technique should be employed. Unfortunately it is rarely supported in statistical packages.

Cause(s) and effect(s)

Questions. *If there are more than two variables which can be labelled 'cause' and 'effect' there are a variety of techniques that can be applied to determine more about the relationship between them. Some of these techniques are simple extensions of ANOVA, regression or correlation analysis while others point towards techniques for data exploration that are covered in the next chapter. Regression is used when there are two or more similar 'effect' variables matched to the same 'cause'. Analysis of covariance (ANCOVA) when two variables are known to be associated and one is used as the dependent variable in an ANOVA analysis. Multiple regression is used when there are*

Chapter 8
The tests 2: tests
to look at
relationships

several 'cause' variables and a single 'effect'. Stepwise regression also has several 'cause' variables and one 'effect' but builds the best fit model in stages. Path analysis is more of a data exploration technique that arranges the interrelationships between several 'causes' and 'effects'.

Regression

If regression has been carried out on several different sets of individuals, in different sites or in different years for example. Then it is possible to compare the slopes of these different analyses to see if they differ. Most statistical packages do not support this type of analysis directly, although a visual comparison of several analyses can be made by plotting the estimated slopes and their confidence intervals to see if they overlap.

ANCOVA (analysis of covariance)

This technique is something of a hybrid between analysis of variance and linear regression. Imagine that a field experiment has been set up with plots given one of five different levels of additional CO_2. Data is collected for sap sugar concentration. This appears to be a simple ANOVA type of design. However, the plots have been surveyed for a range of physical parameters and are known to have a range of organic material in the soil and this affects the sap sugar concentration. The ANCOVA test will effectively use a regression analysis to remove the effect of the organic material level *before* the standard ANOVA is attempted. The factor accounted for by the regression is called the *covariate*, hence the name analysis of covariance. ANCOVA is supported by both MINITAB and SPSS.

An example

A laboratory investigation into the physiological differences between two species of amphipod (shrimp) has shown that although females of both species push water across their developing eggs they do so with different beat frequencies. A further investigation has been made using video cameras in the field to see if this relationship still holds. The investigators also know that beat frequency for both species increases with water temperature and this has been recorded.

Species	Water temperature	Beats/minute
1	10.1	89.0
1	12.2	94.8
1	13.5	99.6
1	11.2	93.8
1	10.2	91.0
1	9.8	89.2

Continued

Chapter 8
The tests 2: tests
to look at
relationships

Species	Water temperature	Beats/minute
2	14.1	104.5
2	12.3	103.6
2	9.5	91.1
2	11.6	99.6
2	10.1	99.1
2	9.4	88.7

Analysis of variance or *t*-test show that the beats/minute is not significantly different between the two species (ANOVA: $F_{1,10}=2.419$, $P>0.1$). However, this does not account for the effect of temperature on beat frequency. An analysis of covariance using species as the grouping variable and temperature as a covariate shows that there is a highly significant difference between the two species ($F_{1,9}=12.436$, $P=0.006$). Note that analysis of covariance has one fewer degrees of freedom for each covariate that is used than an analysis of variance on the same data.

SPSS

SPSS. Input the data in columns, as in the example. Label the columns appropriately. There are at least two methods to reach analysis of covariance, I only consider one here in detail.

Method 1. From the 'Statistics' menu select 'ANOVA models' then 'Simple factorial...'. Move the observed data ('beats_m' in the example) to the 'Dependent:' box. Move the grouping variable ('species' in the example) to the 'Factor(s):' box. 'Define range..' for the grouping variable (see one-way ANOVA, page 86 if you are unsure how to do this). Move the covariate ('temp' in the example) to the 'Covariate(s):' box.

```
* * *  A N A L Y S I S   O F   V A R I A N C E  * * *

              BEATS_M
        by    SPECIES
        with  TEMP

              UNIQUE sums of squares
              All effects entered simultaneously

                              Sum of              Mean              Sig
Source of Variation           Squares     DF      Square      F     of F

Covariates                    241.578     1       241.578   42.180  .000
    TEMP                      241.578     1       241.578   42.180  .000

Main Effects                   71.218     1        71.218   12.435  .006
    SPECIES                    71.218     1        71.218   12.435  .006

Explained                     312.483     2       156.242   27.280  .000

Residual                       51.546     9         5.727

Total                         364.029    11        33.094


12 cases were processed.
0 cases (.0 pct) were missing.
```

Data output 8.14

Chapter 8
The tests 2: tests
to look at
relationships

This confirms that the data for 'beats_m' have been grouped using 'species' with 'temp' as a covariate. The 'UNIQUE sums of squares' and 'All effects entered simultaneously' confirm standard calculation options that you will not want to change until you move on to very advanced statistics. The rest of the output is designed to cope with many factors and therefore has extra lines that appear totally superfluous for this simplest possible scenario. The first 'Source of Variation' is the effect of the covariate then the 'Main Effect' of 'species'. The extra lines are totals but in the example there is only one covariate and one main effect so the total is the same as the single line under 'covariates' and 'main effects'.

The first column is 'sum of squares', then 'DF' or degrees of freedom, then 'mean square' (the mean square value is the sum of square value divided by the degrees of freedom). As there are two species there is one degree of freedom for 'species', the covariate has one degree of freedom. There were six observations within each species giving 11 degrees of freedom for 'Total'.

Finally comes the important bit; the F ratio, labelled 'F' here. This is the mean square for 'species' or 'temp' divided by that for 'residual'. SPSS gives you the P value associated with this value of 'F' and these degrees of freedom and labels it 'Sig of F'. In biology we usually look for a value less than 0.05. Here, the probability is 0.006 for 'species' and indicates that the 'beat frequency' for the two species are highly significantly different from each other.

Method 2. From the 'Statistics' menu select 'ANOVA models' then 'General factorial...' rather than 'Simple factorial...'.

MINITAB

MINITAB. Analysis of covariance is very simple to carry out in MINITAB. Input the data in columns, exactly as in the example and label appropriately. From the 'Stat' menu select 'ANOVA' then 'Analysis of covariance...'. Move the observed data ('beats/m' in the example) into the 'Response(s):' box, the grouping variable ('species' in the example) into the 'Model:' box and the covariate ('temp' in the example) into the 'Covariate(s):' box. Click 'OK'. (Or type 'ANCOVA 'beats/m' = 'species';' at the MTB> prompt and 'covariates 'temp'.' at the SUBC> prompt.)

```
Analysis of Covariance (Orthogonal Designs)

Factor    Levels Values
Species      2    1    2

Analysis of Covariance for Beats/m

Source       DF      ADJ SS        MS      F       P
Covariates    1      241.21    241.21  41.94   0.000
Species       1       71.05     71.05  12.35   0.007
Error         9       51.76      5.75
Total        11      364.03


Covariate    Coeff    Stdev  t-value        P
Temp         2.958    0.457    6.476    0.000
```

Data output 8.15

Chapter 8
The tests 2: tests
to look at
relationships

The output first confirms the test. Then lists the factors (grouping variables) giving the number of groups (labelled 'levels') and the codes assigned to the groups (labelled 'values').

Then comes another confirmation of the test followed by a simple ANOVA table with an extra row for the covariates. The table has degrees of freedom ('DF'), sum of squares ('ADJ SS'), mean square ('MS' which is SS/DF), the *F* ratio ('F' which is factor MS/error MS) and finally the *P* value ('P'). If the *P* value is less than 0.05 then the null hypothesis is rejected. In the example the effect of temperature as a covariate is confirmed to be highly significant 'P' is given as '0.000' which should be reported as $P < 0.001$. The factor 'species' has a *P* value of '0.007' so the null hypothesis that species have the same 'beats_m' is rejected and the alternative hypothesis that the two species differ is accepted.

Finally there is a regression style analysis of the covariate ('temp' in the example). It shows that the relationship has a positive slope ('Coeff' can be thought of as the slope of the relationship and is '2.958' in the example). The probability that the slope is zero is very small, given as '0.000' in the example or $P < 0.001$.

Multiple regression

If there are several 'cause' variables set or chosen by the experimenter and a single 'effect' variable then multiple regression may be appropriate. The same assumptions as for linear regression apply so each of the 'cause' variables must be measured without error. There is an additional assumption that each of the 'cause' variables must be independent of each other. Multiple regression works in exactly the same way as linear regression only the best-fit line is made up of a separate 'slope' for each of the 'cause' variables. There is still a single 'intercept' which is the value of the 'effect' variable when all the 'cause' variables are zero.

A multiple regression using just two 'predictor' or 'cause' variables is possible to visualize using a three-dimensional scatter plot but if there are any more 'predictors' there is no way to satisfactorily display the relationship.

The technique of multiple regression is rather overused as it is supported and fairly easily accessible in most statistical packages. There is the implicit assumption that all the 'cause' variables impinge directly on the 'effect' variable which may appear uncomfortable or unreasonable in certain circumstances. Multiple regression can be found in both SPSS and MINITAB.

Stepwise regression

The assumptions and conditions of stepwise regression are identical to multiple regression. The difference is in the way the best fit model is generated. In stepwise regression the 'causes' are added and subtracted in steps only using those combinations and slopes that generate a better fit (i.e. smaller residuals). This technique is useful as it will help to identify

Chapter 8
The tests 2: tests
to look at
relationships

those 'cause' variables that are most important, which can lead to better experimental design in the future.

Path analysis

This is a technique related to regression and correlation that removes much of the 'cause' and 'effect' baggage. The idea is to generate a map of the inter-relationships between variables to visualize the effects and associations within groups of variables. Unfortunately, this analysis is not supported in most statistical packages.

9: The tests 3: tests for data exploration

Types of data

Data exploration can be attempted with almost any type of data although in practice it becomes most useful where there are a large number of variables and observations. This is because the main aim of data exploration techniques is to synthesize and process the inter-relationships between observations in such a way to make the patterns obvious to the experimenter. Of all the areas of statistics covered in this book this is the one where there is the most scope for personal choice. These techniques are less concerned with *P* values and should be treated as ways to generate hypotheses rather than test them.

Several of the more commonly employed techniques are considered here. This is certainly not an exhaustive list but it does provide a flavour of the sort of techniques that are available.

Observation, inspection and plotting

The simplest and perhaps most obvious way to generate new hypotheses or to explore relationships between variables is to plot them. Many statistical packages, including SPSS and MINITAB will produce a matrix of scatter plots with each cell of the matrix having a different plot of two variables. This sort of visual aid will give a general feel for which variables are related to which as well as for the 'shape' of the data. Don't be afraid to experiment with different types of plot before moving on to more standard methods.

Principal component analysis (PCA) and factor analysis

These are two very similar techniques which weight all the available variables to provide the maximum discrimination between individuals. The idea of principal component analysis (factor analysis, principal axes) is very similar, in many ways, to correlation and regression. The technique can be applied to any data set that has two or more observations for each individual (e.g. several different morphometric measurements from the same specimen). There are assumptions about the data: that it is continuous and normally distributed but these can be overlooked if the purpose of the test is to generate further hypotheses.

The technique can be visualized well when there are only two observations for each individual. First, imagine a scatter plot of two variables that are correlated such that the points fall within an oval cloud. PCA will determine the line through the points that passes through the long axis of the cloud and will use that as the first principal axis or 'principal component'.

A line through the cloud of points at right angles to the first axis will generate a second principal component. Of course, this process occurs in multi-dimensional space within the computer with one dimension for each of the variables included in the analysis and the 'line' through the clouds of points being formed by weighting each of the variables appropriately.

In this way PCA synthesizes the data from a mass of variables into a set of compound axes. The first axis will explain the most variation, then the second and so on. Therefore inspection of the weightings of the first few axes will show which variables contribute most to the differences between individuals.

In morphometric analysis it is usually the case that individual specimens will vary in size. The first principal axis will nearly always account for size and it is often employed as a method for removing size from the analysis leaving aspects of 'shape' for the second and subsequent axes.

An example

A typical use of PCA and factor analysis is for exploration of morphometric characters. Here there is a very small data set from only sixteen individuals of two species of fruit fly. Five morphological characters have been measured to the nearest 0.01 mm, the sex and species is recorded too. The species are coded '1' and '2' for convenience and sex is coded '1' for female '2' for male.

Sex	Species	Thorax length	Wing length	Femur length	Eye width	Third ant. segment
1	1	1.01	2.51	0.06	0.52	0.11
1	1	0.98	2.45	0.05	0.53	0.12
1	1	1.02	2.57	0.08	0.55	0.11
1	1	1.05	2.61	0.07	0.52	0.10
2	1	0.98	2.40	0.04	0.54	0.13
2	1	0.89	2.35	0.04	0.50	0.14
2	1	0.89	2.38	0.05	0.50	0.12
2	1	0.95	2.41	0.05	0.49	0.12
1	2	1.20	3.10	0.09	0.48	0.09
1	2	1.15	3.12	0.10	0.52	0.10
1	2	1.18	3.21	0.09	0.52	0.11
1	2	1.21	3.20	0.10	0.55	0.09

Continued

Sex	Species	Thorax length	Wing length	Femur length	Eye width	Third ant. segment
2	2	0.95	2.51	0.08	0.56	0.11
2	2	0.94	2.50	0.07	0.49	0.13
2	2	0.96	2.62	0.08	0.51	0.14
2	2	0.91	2.45	0.07	0.52	0.13

It is important to realize that there is no requirement for the grouping variables 'sex' and 'species' as the PCA operates on the measured variables to maximize the differences between individuals (i.e. the rows in this data set). I have used the coded variables only to illustrate the results of the analysis.

There are three main components to the output. The first is the weighting applied to each of the variables to generate the principal axes. The second is the set of eigenvalues that show how important the principal axes are (each axis will explain less of the variation than the last). The third is the position of the individuals on the axes, it is this that is used to generate the graphical display of the output (Fig. 9.1).

Investigation of the factor weightings will show which characters are being used to generate the differences between individuals and which are not (the ones with weightings near zero). In the example the first axis (using either PCA or factor analysis) is generated by contrasting the three size variables 'wing', 'thorax' and 'femur' against 'third ant. segment'. This allows the investigator to focus on these characters as the most important. The position of individuals

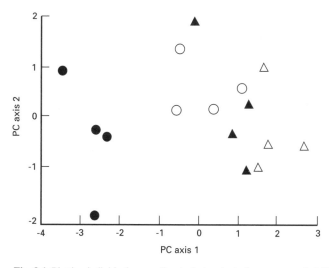

Fig. 9.1 Plotting individuals according to their principal components in MINITAB. PCA maximizes the difference between individuals rather than groups. The individuals are coded by species (shapes) and sex (filled or open shapes). The filled circles are clearly different to the other groups on axis 1 but there is no obvious pattern on axis 2. In many PCA analyses on morphological data axis 1 is closely related to size and other axes to 'shape'.

on the axes can be used in ANOVA to see if the groups vary. In the example a two-way ANOVA on the first principal component (PC1) using 'sex' and 'species' as grouping variables proved that both factors were highly significant and that they had a significant interaction too. A similar test on PC2 showed no significant differences at all. This shows that there are morphological differences between the sexes and species that are condensed into the first axis.

SPSS

SPSS. Input the data in the same form as the table in the example. Label the columns appropriately. From the 'Statistics' menu, select 'Data reduction' then 'Factor…'. In the dialogue box highlight all the measured variables (but not any grouping variables) and move them into the 'Variables:' box. You can click 'OK' now although there are two detours that may pay dividends. Clicking on 'Extraction' will allow you to determine the number of axes ('Factors:') that are generated (I selected two for this example). Clicking on 'Scores' allows you to store the scores as variables. If you do this then you can generate a figure similar to that I have from the example data by using a scatterplot. (Fig. 9.1)

The output generated is as follows:

```
- - - - - - - - - - -   F A C T O R   A N A L Y S I S   - - - - - - - - - - -

Initial Statistics:

Variable      Communality  *  Factor   Eigenvalue   Pct of Var   Cum Pct
                           *
THORAX           1.00000   *    1        3.41932        68.4        68.4
WING             1.00000   *    2         .98827        19.8        88.2
FEMUR            1.00000   *    3         .38289         7.7        95.8
EYE              1.00000   *    4         .19284         3.9        99.7
THIRD_AN         1.00000   *    5         .01668          .3       100.0

PC    extracted   2 factors.

Factor Matrix:

                 Factor  1      Factor  2

THORAX            .95896        -.11237
WING              .95234        -.17829
FEMUR             .89104         .01764
EYE               .22859         .97026
THIRD_AN         -.86403        -.04634


Final Statistics:

Variable      Communality  *  Factor   Eigenvalue   Pct of Var   Cum Pct
                           *
THORAX            .93223   *    1        3.41932        68.4        68.4
WING              .93874   *    2         .98827        19.8        88.2
FEMUR             .79427   *
EYE               .99365   *
THIRD_AN          .74870   *

- - - - - - - - - - -   F A C T O R   A N A L Y S I S   - - - - - - - - - - -

Skipping  rotation   1 for extraction   1 in analysis  1

 2 PC  EXACT  factor scores will be saved.

Following factor scores will be added to the working file:

 Name       Label

FAC1_2      REGR factor score   1 for analysis   1
FAC2_2      REGR factor score   2 for analysis   1
```

Data output 9.1

A lot of the output is unimportant or rather repetitive. There are two sections to focus on. First the 'Factor matrix:' section that gives the weightings for the measured variables for the two factors. In this case 'Factor 1' shows a high positive weighting for the first three variables contrasted against a high negative weighting for 'Third_an' whereas 'Factor 2' gives a high positive weight to 'eye'.

The second important section is the right-hand side of the 'Final statistics:' section. The relative importance of the two factors is given here first as an 'Eigenvalue' and then converted into a percentage of the variation 'Pct of Var' and cumulative percentage of variation 'Cum Pct'. In the example the first factor accounts for 68.4% of the variation and the second 19.8%.

MINITAB

MINITAB. Input the data in columns as set out in the example. Label all the columns appropriately. From the 'Stat' menu choose 'Multivariate' then 'Principal components...'. In the dialogue box highlight all the measured variables then click on 'Select'. The analysis will calculate as many components as there are variables (i.e. five in the example) unless you choose a lower number in the 'Number of components to compute:' box. If you wish to keep the position of individuals on each of the axes then you must choose columns for values in the 'Scores:' box. In the example I selected two components and then stored them in columns c9 and c10, then scatterplotted c9 and c10 to produce the figure.

If you click OK this is the output you will get:

```
Principal Component Analysis

Eigenanalysis of the Correlation Matrix

Eigenvalue    3.4193    0.9883    0.3829    0.1928    0.0167
Proportion    0.684     0.198     0.077     0.039     0.003
Cumulative    0.684     0.882     0.958     0.997     1.000

Variable         PC1       PC2
thorax        -0.519    -0.113
wing          -0.515    -0.179
femur         -0.482     0.018
eye           -0.124     0.976
third_an       0.467    -0.047
```

Data output 9.2

The output confirms the test. Then it gives the eigenvalues of the principal components from one to five. The eigenvalue can be treated as a measure of the variation explained by the axis. Each factor will explain less of the variation than the previous one. This is converted to a proportion of the variation and cumulative proportion for convenience. So the first axis explains 68.4% of the variation and the second 19.8%.

In the second part of the output the weightings applied to the five measured variables is given. The first principal component can be interpreted as a contrast between 'third_an' and 'thorax', 'wing' and 'femur' while the second is heavily influenced by 'eye'. Remember these weightings maximize the differences between individuals.

Excel. It is not possible to carry out this test in Excel.

Canonical variate analysis (CVA)

This technique works in very much the same way as PCA but with one crucial difference the individuals must be assigned to groups before the analysis is run. The test then calculates the variable weightings that will maximize the differences between *groups* rather than individuals, as is the case with PCA.

Canonical variate analysis produces weightings that will allow you to identify those variables that are the most different between groups and discard those that are the same. It is important to use real classifications rather than arbitrary ones to get the maximum benefit from the technique.

Discriminant function analysis

As with CVA, this technique also requires that the individuals be divided into groups. The idea of discriminant function analysis is to provide a set of weightings that allow the groups to be distinguished. The weightings can then be used on individuals that are not assigned to a group to provide a probability of them belonging to each of the possible groups. If the probability is high then the 'unknown' can be confidently assigned to a group. The power of the weightings is often tested by removing a portion (one or more) of the data set, using the remainder to create the weightings and then to use the weightings to assign the removed individuals to groups. The hit rate is a measure of the power of the test to discriminate real unknowns.

An example

Here we shall be using the same example as for the PCA but using one of the grouping variables as the target discriminator. The analysis will determine how often an individual of unknown species or sex would be attributed to the correct group.

SPSS

SPSS. Input the data in columns and label appropriately. From the 'Statistic' menu select 'Classify' then 'Discriminant...'. Move the variable with the group codes into the 'Grouping variable:' box. Click on 'Define range...' to input the lowest and highest group code numbers. Click 'Continue'. Next move all the measured variables into the 'Independents:' box. There are now a range of possible options that can be selected from the buttons at the bottom. The most useful is 'Save...' where the 'Predicted group membership' and the probabilities of belonging to a particular group can be stored as separate variables. If you don't do this then there is no way of checking the accuracy rate of the discriminant analysis. Another important option to consider is under the 'Classify...' button where you can either allow the probability of group membership to be equal or to be determined by the relative size of the groups. In the example the groups are of equal size anyway so there is no difference but this will not always be the case. Click 'OK'.

Masses of output appears in the '!Output' window of which this is an edited section.

```
- - - - - - - -   D I S C R I M I N A N T   A N A L Y S I S   - - - - - - - -

On groups defined by sex

                    Canonical Discriminant Functions

                  Pct of   Cum  Canonical  After  Wilks'
    Fcn Eigenvalue Variance  Pct    Corr     Fcn  Lambda  Chi-square  df  Sig

                                           :    0 .273675    14.902    5  .0108
    1*    2.6540  100.00 100.00   .8522 :

       * Marks the 1 canonical discriminant functions remaining in the analysis.


    Standardized canonical discriminant function coefficients

                    Func  1

    thorax       2.07577
    wing        -1.65920
    femur         .16962
    eye           .09039
    third_an     -.26077


    Structure matrix:

    Pooled within-groups correlations between discriminating variables
                                  and canonical discriminant functions
    (Variables ordered by size of correlation within function)

                    Func  1

    thorax        .77039
    third_an     -.74360
    wing          .52341
    femur         .37016
    eye           .13923


    Canonical discriminant functions evaluated at group means (group centroids)

       Group     Func   1

         1        1.52389
         2       -1.52389
```

Data output 9.3

Most of the output is rather repetitive and some of the most interesting bits of the output only appear on the spreadsheet itself (if you selected options in the 'Save...' box). The sections to focus on are the coefficients. In the standardized coefficients the weightings given to each of the measured variables to maximize the differences between groups are given. Clearly 'thorax' and 'wing' are the highest weighted in the example. The structure matrix gives a slightly different view of the data as it appears that although 'thorax' has the highest correlation with the discriminating function the negative correlation for 'third_an' is almost as important.

Finally comes the 'group centroids' for the two groups on the axis of the discriminant function. As there are only two groups and they are of equal size in the example it is not surprising that the group centroid (mean position) are the same distance either side of zero. In practice this means that any individual that has the weightings applied to the measurement variables and scores more than zero will be assigned to group 1 and less than zero to group 2.

The group assignments are made in a new variable on the spread sheet. In the example all individuals were assigned to the correct group apart from the individual in row 5.

MINITAB

MINITAB. Input the data into columns. Make sure that the grouping variable is coded as an integer (e.g. female = '1' : male = '2'). Label the columns appropriately. From the 'Stat' menu select 'Multivariate' then 'Discriminant analysis...'. Move the grouping variable ('sex' or 'species' in the example) into the 'Groups:' box. Highlight all the measured variables and click 'Select' to move them into the 'Predictors:' box. Click 'OK'. (In the example I chose 'sex' as the grouping variable as the analysis was 100% accurate when 'species' was used.)

You will get the following output:

```
Discriminant Analysis

Linear Method for Response:    sex
Predictors:   thorax   wing   femur   eye   third_an

Group           1          2
Count           8          8

Summary of Classification

Put into       ....True Group....
Group            1          2
1                8          1
2                0          7
Total N          8          8
N Correct        8          7
Proport.      1.000      0.875

N =    16     N Correct =    15     Prop. Correct = 0.937

Squared Distance Between Groups
                 1          2
1           0.00000    9.28891
2           9.28891    0.00000

            Linear Discriminant Function for Group
                 1          2
Constant      -795.5     -759.6
thorax         927.6      838.2
wing          -117.7      -97.2
femur         -264.2     -293.4
eye           1155.6     1143.9
third_an      3098.6     3174.4

Summary of Misclassified Observations

Observation      True      Pred     Group Sqrd Distnc Probability
                 Group     Group
       5 **        2         1          1      9.183      0.524
                                        2      9.378      0.476
```

Data output 9.4

The output confirms the test and then the variables used for grouping and prediction. It then gives the number of observations in each of the groups (in the example there were eight females and eight males). Next comes a 'Summary of Classification' where the real group 'True group' and the group that each individual is put into is compared. In the example all of group 1 were correctly assigned to group 1 but one individual from group 2 was incorrectly assigned to group 1. A summary of the accuracy comes next giving it as a proportion correct of 0.937 in the example (i.e. 93.7% or 15/16 of the individuals were assigned to the correct group). The squared difference between the groups is not very useful for just two groups but will show which groups are most similar to which if there are more than two. Then comes a section labelled 'Linear discriminant function' which gives weightings to each of the measured variables that are akin to the slopes in a multiple regression. Finally there is a summary of the misclassifications. In the example only one of the 16 was misclassified. The individual was in row 5. The analysis gave it a 52.4% probability of being in group 1 (i.e. wrong) and 47.6% of group 2 (i.e. correct) so it was a borderline case.

The analysis suggests that an unknown individual from this area could be assigned to 'sex' from analysis of the measured variables alone with an accuracy of 93.7%.

Excel

Excel. You cannot carry out this test in Excel.

Multivariate analysis of variance (MANOVA)

All the analysis of variance techniques that are introduced in Chapter 7 have only one observed variable although they had one or more classification variables. Therefore they can all be called *univariate* analyses. The statistical analysis MANOVA allows more than one observed variable to be analysed at once.

Multivariate analysis of covariance (MANCOVA)

This technique is related to MANOVA in the same way that ANCOVA is related to ANOVA. If there are more than one observed variable, one or more ways of classifying the data and furthermore there is a measured observation that is known to have an effect on the observations then MANCOVA can be used to remove the effect of this confounding variable from the analysis.

Cluster analysis

This is a general term for a huge range of techniques for the classification of individuals. These techniques are becoming increasingly important as more detailed statistical analysis of DNA sequence analysis is being attempted. Cluster analysis can be used to generate dendrograms that show putative phylogenetic relationships, or at least divide individuals into groups that might have taxonomic meaning. However, cluster analysis is certainly not restricted to molecular sequence analysis, it has a long history in taxonomy (both cladistics and phenetics) and in community ecology (particularly in vegetation classification or ordination).

In its simplest form cluster analysis can be imagined as a step-by-step process. First the individuals are depicted as a scatter of points. Then the individuals that are closest together are identified and their similarity recorded as the distance between them (this is sometimes called a Mahalanobis distance). The two points are amalgamated into a single point half-way between the two. The next two closest points are then identified and amalgamated. This process continues until there is only one point.

SPSS

SPSS. There are several methods available. Input the data in columns so that each individual is represented by a row. From the 'Statistics' menu select 'Classify' and then choose one of the options offered. Simple dendrograms can be generated by selecting the submenu 'plots' in the 'Hierarchical cluster…' option.

MINITAB

MINITAB. There is a variety of clustering methods available in MINITAB. Input the data in columns so that each individual is represented by a single row. From the 'Stat' menu select 'Multivariate' and then one of the three 'Cluster' methods. To generate the dendrogram shown in Fig. 9.2, I used the 'Cluster observations…' method on the measured variables from the example data for PCA.

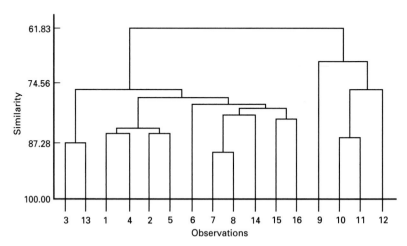

Fig. 9.2 A typical dendrogram showing the results of a cluster analysis in MINITAB. This figure was generated using the sixteen individuals used as the example data for PCA. There are some clear groupings. For example, individuals 9–12 form a separate group on the right of the diagram.

[186]

Excel. No clustering techniques are available in Excel.

DECORANA and TWINSPAN

DECORANA or DCA (detrended correspondence analysis) and TWINSPAN (two-way indicator species analysis) are two analyses developed at the Institute for Terrestrial Ecology which are now widely used in the comparison of communities from species abundance data and in the exploration of ecological data.

10: Symbols and letters used in statistics

One of the surest ways of making a statistics book difficult to read is the tendency to use Greek letters, single italicized letters or obscure symbols. As far as possible I have tried to avoid this in this book. Here are the ones that you are most likely to encounter.

Greek letters

These are often used to signify the true values of particular statistics (i.e. the value you would get if you were able to measure the entire population rather than a sample). The estimates you get of the true values are often then labelled with the corresponding normal letter.

Π	(pi) product of the terms following it (multiply together)
π	(pi) a constant (3.142) used in geometry
Σ	(sigma) sum of the terms following it (add up)
α	(alpha) the critical significance level for the rejection of a hypothesis (usually 0.05)
β	(beta) true regression coefficient (estimated by the statistic, b)
χ	(chi) χ^2 is a commonly encountered statistical distribution
γ	(gamma) γ_1 is the true value of skewness γ_2 is the true value of kurtosis
μ	(mu) true mean of a population
ρ	(rho) true correlation coefficient (estimated by the statistic, r)
σ	(sigma) the true standard deviation of a population
σ^2	(sigma squared) the true variance of a population
T	(tau) the statistic of Kendall's rank-order correlation
Δ or δ	(delta) increment (tiny difference or change)

Symbols

$^{-}$	(overbar) indicates a mean
$\sqrt{}$	square root
$=$	is equal to
$\equiv$	is identically equal to
$\neq$	is not equal to
$\approx$	is approximately equal to
$\sim$	is distributed as
$\|\ \|$	absolute value of the number between the bars. $\|-6\|=6$
$!$	factorial (e.g. $3!=1\times2\times3=6$)
$\pm$	plus or minus (so 5 ± 2 means a range from 3 to 7)
$\geq$	is greater than or equal to

<	is less than (points to smaller value)
<<	is much less than
>	is greater than (points to smaller value)
>>	is much greater than
^	used in some statistical packages (e.g. Excel) to mean 'raise to the power of'
$\cap$; $\cup$; $\subset$; $\not\subset$	symbols used in set work (intersection; union; is a subset of; is not a subset of)
*	indicating a significant result (usually a *P*-value is flagged at <0.05)
*	used in statistical packages to mean 'multiplied by'
**	denotes a highly significant result (usually $P<0.01$)
**	used in some statistical packages (e.g. SPSS) to mean 'raise to the power of'
***	denotes a very highly significant result (usually $P<0.001$)
___	used to underline groups that are not significantly different (see *post hoc* tests, page 108)
∞	infinity (an infinite number) often used in statistical tables to indicate the value of an asymptote
$\propto$	varies directly as, is proportional to
$\Rightarrow$	implies
$\times$ or .	multiply
$\therefore$	therefore

Upper case letters

CI	confidence interval
CL	confidence limit
CV	coefficient of variation
DF	degrees of freedom (also d.f. or df)
F	*F* value (e.g. the output from ANOVA) ratio of within and between group variance
F	sometimes used to indicate a function
H_0	null hypothesis (the uninteresting hypothesis — nothing is happening)
H_1	alternative hypothesis (the interesting hypothesis — something is happening)
MS	mean square (SS/d.f. in an ANOVA table)
P	probability
PI	prediction interval
SD	standard deviation (also *s*)
SE	standard error
SS	sum of squares
X^2	estimate of value for χ^2 (chi-square)

Lower case letters

a	the intercept of a regression line (where the line crosses the y-axis)
b	slope of a regression line
df	degrees of freedom (sometimes d.f. or DF)
e	a constant ($=2.172$) used as the base for natural or Naperian logarithms (ln)
g	estimate of value of γ (gamma), $g_1=$ skew, $g_2=$ kurtosis
f	used to indicate a function
i	often used to indicate a sequence of observations (e.g. x_i)
j	often used to indicate a second sequence of observations (e.g. x_{ij})
m	often used to indicate the sample mean
p	binomial probability (e.g. 0.5 probability of an individual being female)
r	measure of correlation (Pearson product moment correlation, varies from -1 to 1)
r_s	measure of correlation produced by Spearman's rank order correlation
r^2	a measure of the amount of variation accounted for by a regression line or correlation
s	standard deviation of a sample (also SD)
s^2	variance of a sample
t	value of the statistic resulting from a Student's t-test
v	occasionally used to indicate variance of a sample
x	often used to indicate an observation
y	often used to indicate a second observation on the same individual as x
z	often used to indicate a third observation on the same individual as x and y

11: Assumptions of the tests

Most statistical tests make assumptions about the data they are being applied to, if the assumptions are violated it is wise to treat the results with caution especially when *P* values fall in the range 0.01 to 0.1.

Here is a test-by-test summary of the assumptions:

Test	Assumptions
G-test	Observations can be assigned to groups or categories
Chi-square test	Observations can be assigned to groups or categories
Kolmogorov–Smirnov	Observations come from a fairly continuous scale
Paired *t*-test	Both sets of data are normally distributed Variance is the same in both samples (although there are tests often incorporated into statistical packages that make corrections)
Wilcoxon's signed ranks test	Observations are made on a scale such that the magnitude of differences is meaningful
Sign test	Observations are made on a scale so that the question 'is A bigger than B?' can be answered
t-test	Both sets of data are normally distributed Variance is the same in both samples (although there are tests often incorporated into statistical packages that make corrections)
Mann–Whitney *U* test	Observations are made on a continuous scale (i.e. they can be put into rank order with very few ties)
Friedman test	One observation per factor combination Observations may be put in meaningful rank order

Test	Assumptions
All ANOVA tests	Observations are independent both within and between samples Variance is the same in all samples Data are normally distributed Observations are assigned to groups (coded by integers) using one or more factors
Kruskal–Wallis test	Observations are made on a fairly continuous scale (i.e. they can be put into rank order with very few ties)
Scheirer–Ray–Hare test	Observations are made on a continuous scale (i.e. they can be put into rank order with very few ties)
Chi-square test of association	Observations can be assigned to categories or groups using one or more factors
Phi coefficient of association	Observations can be assigned to two groups for each of two factors
Cramér coefficient of association	Observations can be assigned to categories or groups using two factors
'Standard' correlation (Pearson product–moment correlation)	Individuals have observations for two variables measured on a continuous scale Two variables are both normally distributed
Spearman's rank-order correlation	Individuals have observations for two variables measured on an approximately continuous scale
Kendall's rank-order correlation	Individuals have observations for two variables measured on an approximately continuous scale
Kendall's robust line-fit method	'Effect' measured on an approximately continuous scale 'Cause' on any meaningful scale
ANCOVA (analysis of covariance)	Observations and covariate measured on a continuous scale Variance the same for all factor levels Observations normally distributed Observation independent

Chapter 11 Assumptions of the tests	Test	Assumptions
	'Standard' regression (model I linear regression)	'Cause' (= independent or x) variable measured without error Variation in 'effect' (= dependent or y) the same for all values of 'cause' Relationship between x and y is linear 'Effect' is measured on a continuous scale
	Logistic regression	'Cause' (= independent or x) variable measured without error Variation in 'effect' (= dependent or y) the same for all values of 'cause' Relationship between x and y is linear 'Effect' is expressed as a proportion
	Model II regression	Individuals have observations for 'y' variable measured on an approximately continuous scale
	Polynomial regression	As standard regression but not assuming that the relationship between x and y is linear
	Stepwise regression/multiple regression/path analysis	As standard regression but with several 'effect' variables measured for each individual
	Discriminant function analysis	Individuals have two or more observations assigned to them measured on continuous scales
	Principal component (Factor) analysis	Individuals have two or more observations assigned to them measured on continuous scales
	Canonical variate analysis	Individuals have two or more observations assigned to them Observations are measured on continuous scales Individuals can be assigned to groups
	MANOVA (= multivariate analysis of variance)	Two or more observations for each individual Observations are independent both within and between samples Observations are assigned to groups (coded by integers) using one or more factors

Test	*Assumptions*
	Variance is the same in all samples
	Observations are normally distributed
MANCOVA (= multivariate analysis of covariance)	Two or more observations for each individual
	Observations are independent both within and between samples
	Variance is the same in all samples
	Observations are normally distributed
	Observations are assigned to groups (coded by integers) using one or more factors
	Covariate is measured on a continuous scale
Cluster analysis (a family of techniques that have slightly different assumptions)	Each individual has two or more observations assigned to it
	Observations are measured on meaningful scales
	Individuals can be assigned to groups

What if the assumptions are violated?

There are several possible courses of action that can be taken (in approximate order of preference):

1 Data could be transformed to make them suitable for the analysis chosen.
2 An alternative test of the same hypothesis but with different assumptions is used instead.
3 The hypothesis is reframed to allow a different test to be used.
4 Violation of the assumptions could be ignored totally but the results regarded with caution.
5 No test is carried out at all.

12: Hints and tips

Using a computer

- Save frequently—packages usually have a little disk icon on screen all the time, click on it! (computers 'hang', 'lock' or 'crash' every now and again and you want to make sure you don't lose data)
- Learn a few keyboard shortcuts
- An easy way to select a block of text or data in many packages is to place the cursor at the beginning, move the pointer to the end and press shift as you left click the mouse
- Using the underlines—the underlined letters in menus mean that you can access the menu by typing the letter on the keyboard while holding the 'alt' key
- Use the 'tab' key to move between boxes—useful in many of the Windows dialogue boxes
- Use 'shift' and 'tab' together to move backwards through boxes—useful to correct mistakes
- Back-up your important files frequently on floppy disks
- Holding 'alt' and pressing 'tab' moves you between open packages—this is extremely useful!
- Edit in the best editing package, then do the statistics or graph drawing in another—do not feel you have to use the pathetic spreadsheet capabilities of the statistics package
- If you are given data in the format of another package that your package cannot read you can nearly always read it by saving in raw text format from the first package
- When converting labels into numbers I suggest that using alphabetical order all the time will avoid many problems of converting the numbers back to labels
- Cut & paste is a very powerful facility of most packages—you can usually copy material from one to another using copy and paste
- The keyboard shortcuts for 'cut', 'copy' and 'paste' are nearly always 'ctrl+x', 'ctrl+c' and 'ctrl+v' respectively. Using the shortcuts is easier and quicker than going to the 'edit' menu and selecting from there
- Double clicking or right clicking often brings up helpful options
- If you get stuck try the help file—these are usually extensive and often have examples too
- 'Alt+F4' will usually close a package
- Don't leave lots of unnecessary windows open—they slow the computer down (e.g. graphs may be in separate windows)

Sampling

- Try to balance sampling designs if you can (i.e. take equal sized samples)
- Measure everything you can easily — you never know what is going to be important
- Avoid sampling at regular intervals if possible
- Choosing the nearest individual to a random point will *always* bias the sample to individuals on the edge of clumps and against those in the middle
- Don't carry out repeat sampling in the same sequence
- Measure to sensible precision only, not to maximum
- Three sub-samples from a site are much better than two
- When setting up a laboratory experiment make sure it is genuinely factorial (i.e. there should be no confounding factors such as 'all species x was from the sunny site')
- Randomize measurement routines as much as possible (i.e. don't measure all of group 1 then 2, etc.)
- Try double-blind labelling if possible (i.e. when measuring you don't know to which group the individual belongs)
- Don't try to design overelaborate experiments — it is difficult to interpret anything with more than three factors
- Use transects with caution
- Always sample with a clear idea of the statistical test you intend to use in mind

Statistics

- Try the analysis on dummy data before collecting any real observations
- Find a design like the one you are doing in a statistics book and try to repeat the result in your statistical package
- Frame null hypotheses very carefully before anything else
- Always consider whether the data violate the assumptions of the test — if they do be wary of the results
- Transformation of the data can often turn an inappropriate data-set into an appropriate one
- One-tailed tests have their place (i.e. the alternative hypothesis is 'x is greater than y' rather than 'x is different to y'), but if in doubt use two-tailed tests
- If P values are close to 0.05 consider resampling to increase sample sizes
- There is nothing 'special' about $P = 0.05$ so don't be completely tied to it
- In regression, if you are unsure which variable is the 'cause' and which is the 'effect' then the data is probably not suitable for regression anyway
- If a nonparametric test is available use it
- Carry out tests on incomplete data sets to get a feel for the results from the complete set
- Use 95% confidence intervals rather than standard errors

Displaying the data

- Never use 3D effects (except, possibly for posters)
- If you must use a 3-axis graph make sure that every point is anchored to the 'floor' by a spike, otherwise there is no way of determining its position on *two* of the axes
- Use the minimum amount of shading
- Use black and white for everything
- Avoid putting titles on graphs and figures
- Use figure legends for every graph and make sure that the legend is informative enough to make the graph intelligible without reading the main text of a report
- Don't use any more decimal places than you have to and, for raw data, no more than you have measured
- If a graph has a measure of position (e.g. mean) then always display a measure of dispersion as well
- If you want the reader to compare figures make sure they have the same scales (especially on the *y*-axis)
- If you use a line graph it must be possible for intermediate values to exist (they are implied by the line!)
- Don't be afraid to use log scales even when the observations are not logged
- Never draw best-fit lines unless the data is suitable for regression
- Never extend best-fit lines beyond the data
- Always have gaps between bars on a bar chart (data is discrete)
- Never have gaps between bars in a histogram (data is continuous)
- Use pie charts for categorical data
- Don't clutter graphs with too much information
- If plotting means always plot error bars too (95% CI or SD)
- If plotting medians always plot quartiles too

Glossary

An explanation of commonly encountered words, concepts and acronyms including thumbnail description of many statistical tests.

a posteriori phrase applied when a hypothesis is generated after the data has been collected. Sometimes used as a synonym for a *post hoc* test

a priori phrase applied when a hypothesis is generated before the data has been collected

Abscissa the x or horizontal axis of a graph

Accuracy closeness of a measure to its true value (different to precision)

Anderson–Darling test a statistical test used to determine whether a set of data is normally distributed

Angular transformation synonym of arcsin transform. A transformation that is often used to 'normalize' percentage data

ANCOVA short term for analysis of covariance

ANOVA short term for analysis of variance, the term was originally coined by Tukey

ANOVAR alternative term for analysis of variance

Arcsin square-root transformation see arcsin transform

Arcsin transform an operation that is often used to 'normalize' percentage data (synonym for arcsin square-root transformation and angular transformation)

Arithmetic mean a measure of position: sum of all values divided by the number of observations (synonym of mean)

Association measure of the strength of the relationship between two variables. Often used as a synonym of correlation

Assumptions many statistical tests assume that the underlying distribution of the measurements of a whole population is of a particular type (usually that it follows a normal distribution) or that measurements are made without error, etc. etc.

Asymptote the value of a line that a curve is approaching but never meets

Attribute when used to describe an observation it usually implies a small number of possible categories that have no meaningful sequence (e.g. 'blue', 'pink' and 'white')

Autocorrelation a common problem with sets of observations is that they are not truly random and are more similar to their 'neighbours' in space or time than is expected by chance

Average a synonym for arithmetic mean (used by Excel)

Back transformation process by which observations are transformed back to their original units

Bar chart a graph to display the distribution of a discrete variable where each category on the x-axis represents one possible value

Bartlett's test a test for homogeneity of variance

Bernoulli distribution values can only be one of two possibilities (e.g. 0 or 1)

Best fit the statistical model which explains the most variation. The best fit is often constrained by the type of statistical test. For example if the model is a linear regression then the 'best fit' will be the straight line which accounts for the most variation (leaves the smallest residual variation)

Between sample variance synonym for between groups sum of squares. In ANOVA and regression a measure of the variation between factor levels. Comparison of the between sample variance and within sample variance is how ANOVA works

Bias any systematic error in measurement

Bimodal a frequency distribution that has two peaks

Binomial distribution a theoretical probability distribution if events can occur in one of two categories. Can be used as a null hypothesis to determine whether given data are random or not

Bivariate tests which are applied to two variables

Bivariate normal an assumption of many bivariate parametric tests is that the two variables are both normally distributed

Bonferroni method a *post hoc* test, used after a one-way ANOVA to determine which groups are different from which

Bootstrapping a method of analysis that improves many estimated statistics by using repeated subsamples of a data set

Box plot synonym of box and whisker

Box and whisker a method of displaying data where a horizontal line represents the median, a box extends to cover the interquartile range and a line may extend away from the box to the extreme values

Calibration curve regression can predict y from any value of x. This can be used to generate a calibration curve

Canonical variate analysis a multivariate test where the data is divided into groups and the test weights variables to maximize differences between the groups

Caption a small piece of text that should accompany *every* figure and table used in a report (synonym for legend)

Categorical description of data or a variable that is only described by categories. This can be as simple as 'red, blue, green' or an observer-imposed scale, e.g. of aggressiveness where 5 is aggressive and 1 is passive. Categories are often called attributes

Causation the assumption in statistical tests that the value of the independent variable causes the result in the dependent variable

Chi-squared distribution a distribution widely used in statistics that is based on the deviation of sample variance from the true variance of a population. A χ^2 distribution can be generated from a population of standard normal deviates as the probability density of a very large number of samples scaled thus: $((n-1)s^2)/\sigma^2$.

Chi-squared goodness of fit a chi-square test where the expected values are taken from a particular distribution against which the observed distribution is being compared

Chi-square test a contingency table-based statistical test to explore hypotheses of association between variables, expected values are generated by the table

Cluster analysis a number of multivariate tests that group observations by similarity and may provide insight into the data

Confidence interval a measure of spread: it uses the standard error and the t-distribution to give a range of values within which there is a percentage probability of the true mean occurring (usually using set at 95%). In regression the 95% confidence intervals of slope and intercept are used to generate 95% confidence intervals of the line. This zone is not parallel with the best fit line but wider at extreme values of x

Confounded design this often occurs in a multifactorial design when not all combinations have been used or when the effects of one factor cannot be disentangled from the effects of another. Often a result of lack of space or resources rather than poor design

Contingency table a way of displaying data that has been assigned to categories for two variables (e.g. broadleaf/conifer and insect damage > 10% yes/no)

Conservative test one where the chance of Type I error is reduced and that of a Type II error increased (i.e. the null hypothesis is rejected less often than it should be)

Constant any fixed value. In regression analysis the intercept (value of y when x is zero) is sometimes called this

Continuity correction a method of correcting bias in various contingency tests. Makes test result more conservative

Continuous a description of data or variables indicating every value is possible (in theory). E.g. any linear measurement, > 29 possible values as a rule of thumb

Control a general term given to the section of an experiment which is unmodified by the experimenter

Cook's distance statistic a useful measure of the influence of single measures on the outcome of regression analysis (any value > 1 is large)

Correlation a method of measuring the association between two variables (often used as a synonym for Pearson product–moment correlation). Warning: large samples will nearly always produce 'significant' results

Covariance technically, this is the sum of the product of the deviations from the mean for a set of paired observations (i.e. the variance that remains once the relationship between two variables is accounted for). Related to variance and useful in many statistics

Cox regression a type of regression used for assessing hazards (e.g. mortalities) to make predictions about the chance of events occurring

Cramér coefficient the result of a test on contingency tables other than 2×2

Cramér's *V* a statistic correcting chi-square values for sample size

Cumulative frequency the number of times an observation takes a particular value *or less* in a data set (e.g. if there were 26 zeroes, 22 ones and 15 twos then the cumulative frequency for two would be $26 + 22 + 15 = 63$)

Cumulative probability the probability of achieving a particular value and all values greater (or smaller). A type of data display where probabilities are accumulated from left to right until all observed values have been included

Data observations recorded during an experiment or survey

Datum a single observation

Degrees of freedom a number related to the sample size that accounts for the number of observations made, number of factor levels and any manipulations carried out

Dependent variable the observed or 'effect' variable not set by the experimenter

Derived a value that results from a combination of two or more values (e.g. ratio, proportion or percentage)

Descriptive statistic anything that summarises the data. E.g. mean or standard deviation

Dichotomous variable a variable that can only take two possible values (e.g. yes/no)

Discontinuous data only a limited number of values are possible (<30 as a rule of thumb)

Discontinuous variable a variable that only has a limited number of values (usually but not always integers)

Discrete a description of data or variables indicating that not every value is possible, e.g. number of children can never be 2.5

Discriminant function analysis a multivariate test that assigns individuals to groups

Dispersion the way in which the data is distributed, often measured by standard deviation (synonym of spread)

Distribution the spread of a set of observations

Double-blind an experiment when the experimenter does not know which treatment is being applied until labels are decoded. A very good way to avoid bias

Dummy data made up before the experiment is carried out, used for a dry run of the statistical tests planned

Dunn–Sidák method a *post hoc* test, used after a one-way ANOVA to determine which groups are different from which

Dunn test one of many *post hoc* tests used in ANOVA to determine which groups are different from which

Error a word quite widely used in statistics. Can be applied to the degree of accuracy with which measurements are made. Can also be used in statistical tests such as ANOVA to refer to variation not accounted for by the model being tested

Estimate any statistics (e.g. mean) calculated from a sample is an estimate of that value for the whole population

Expected frequency the occurrence of an event determined from a theoretical distribution or hypothesis based on previous observations

Expected value values needed in contingency tables when they are calculated using the row and column totals or values derived from a theoretical distribution or hypothesis

Experiment random assignment of subjects to controlled 'experimental' conditions

Experimental design a process of planning which should always occur before an experiment begins to maximise the usefulness of results obtained while minimizing the effort

Exponential distribution a distribution that arises when there is a constant probability of an event occurring (e.g. birth rate or radioactive decay rate). Logarithmic transformations will turn exponential curves into straight lines

Extrapolation whenever predictions are made beyond the range of the data available

F **distribution** an asymmetric, continuous distribution, used in ANOVA tests with a modal value of 1, representing the frequency of occurrence of the ratio of between group variance/within group variance for groups with the same distribution. Called *F* in honour of R.A. Fisher.

*F***-test** a statistical test with the null hypothesis that groups have the same variance

F **ratio** synonym for *F* value

F **value** the output from an ANOVA test. The ratio of between group variance and within group variance

Factor a grouping variable or independent variable. For ANOVA factors must be discontinuous/categorical or made to appear so

Factor level a number representing different groups, e.g. sap flows, are examined from three species of tree. The species being groups that are arbitrarily assigned numbers as factor levels

Factorial (experiment) an experiment or survey where there are two or more factors and each possible combination of factor levels is represented in the data

Field experiment a general term given to any experiment where conditions are manipulated by the experimenter which occurs outside a controlled laboratory

Figure any graphical item in a report

First-order interaction an interaction between two factors in a factorial ANOVA

Fisher's exact test a statistic for 2×2 contingency tables used when the total number of observations is small

Fisher's *z* transformation a method for transforming the Pearson's correlation coefficient to make it more amenable to analysis

Fixed effect this is often applied to describe a factor in ANOVA that is set by the experimenter (in contrast to a random effect)

Four-way… see 'two-way…' and extrapolate

Friedman test a nonparametric test that is a restrictive version of a two-way ANOVA. Only one observation is allowed for each factor combination

Frequency the number of times a value (or range of values) occurs in a sample (e.g. there were 56 white-eyed flies in the vial)

Frequency distribution the number of times each possible value occurs in a sample

*G***-test** a form of contingency table test. A.k.a. log-likelihood ratio test

Gamma distribution a family of continuous distributions

Geometric mean a measure of position: the antilog of the mean of the logs of each value

GLIM a statistical package

Goodness of fit any statistic that compares the actual distribution of a variable with a theoretical one

Graphical representation any method of displaying data visually

Group a set of observations with something in common. E.g. all from the same tree

Grouping variable a variable by which observations can be divided into groups. E.g. tree number

Harmonic mean a measure of position: the reciprocal of the mean of the reciprocals of each value

Heteroscedasticity when different groups have unequal variance and therefore the data violates the assumptions of ANOVA and regression

Hierarchical ANOVA synonym of nested ANOVA

Highly significant where the null hypothesis is rejected because the *P* value is much less than the 0.05 level. The actual level varies with authors but *P* must be less than 0.01

Histogram a graph to display the distribution of a continuous variable where each category on the *x*-axis represents a range of values

Homogeneity of variances an assumption of many parametric tests such as the *t*-test or ANOVA is that the variances in the groups are equal. This can be tested using the Levene test

Homoscedasticity when different groups have equal variance and therefore conform to one of the assumptions of ANOVA and regression

Hypothesis this is what is being tested when a statistical test is carried out. For example, 'male and female tree frogs have a different mean weight'

Independence an assumption of many statistical tests. The data collected in a sample is not affected by whether another event has occurred. Many statistical tests are rendered invalid because of lack of independence

Independent samples no individuals measured in sample A are also measured in sample B. If this is violated, the sin is called pseudoreplication

Independent samples *t*-test *t*-test comparing two sets of data that are not paired

Independent variable the 'cause' variable set or varied by the experimenter

Individual in statistics an 'individual' can be a pair, bone, region, species or any number of different things. The 'individual' will provide one observation for each variable under consideration and when the data is arranged in the package will provide a single row

Integer a whole number

Interaction in ANOVA this a measure of whether two or more grouping variables have an additive (no interaction) effect or not (interaction)

Intercept in regression analysis; the point on the *y*-axis when the value of *x* is zero

Interquartile range a measure of spread: if the data are in rank order it uses the range from the value 25% down the list to that 75% down

Interval data when observations are made on a meaningful measurement scale

Jack-knifing a technique to determine bias in statistics by recalculating using a subset of the data

Kendall rank correlation a nonparametric test to measure correlation (association)

Kendall's robust line fit a nonparametric version of regression (rarely used)

Kolmorogov–Smirnov test a nonparametric test that compares two distributions. Very useful for goodness of fit tests and more powerful and convenient than a chi-square goodness of fit when samples are large

Kruskal–Wallis test a nonparametric version of a one-way ANOVA that tests the null hypothesis that two or more groups come from populations with the same median

Kurtosis a measure of the shape of a distribution (sometimes called g_2)

Latin square an experiment set up that reduces possible bias caused by unquantified variation in the environment across an experiment

Least significant difference test the simplest and probably the most widely used *post hoc* test used after a one-way ANOVA to determine which groups differ from which

Legend a small section of text that should accompany every figure or table making it interpretable without recourse to the main text (synonym for caption)

Leptokurtic a distribution that has a lot of observations around the mean and in the tails but fewer in the 'shoulders'

Level a particular treatment; a defined condition for the independent variable (set by the experimenter)

Levene test a test for homogeneity of variances. Used for checking data to see if a parametric test such as ANOVA is appropriate

Liberal test a statistical test where the null hypothesis is rejected more often than it should be

Line chart (graph) a graphical representation of data where points are joined. This makes the assumption that values between points are possible and likely to be fairly represented by the position of the line

Linear regression regression that assumes the 'cause and effect' relationship is best described by a straight line

Log transformation an operation where each observation is logged (usually to base 10)

Logistic regression a version of regression where the 'effect' variable is transformed using logits. Useful when using regression to predict values that have a restricted range of possible values (e.g. proportions or percentages)

Logit transformation used to convert values on a scale with limits (e.g. proportions) to a limitless one. Transformation for a proportion, *x*: Logit $x = \ln(x/(1-x))$

LSD test least significant difference test, a commonly used *post hoc* test, used after a one-way ANOVA to determine which groups are different from which

Mann–Whitney *U* test a nonparametric test of a null hypothesis that two groups come from the same distribution. Synonym of Wilcoxon–Mann–Whitney test and Wilcoxon's signed ranks test.

MANOVA acronym for multivariate analysis of variance. A test where there is more than one dependent variable under investigation.

MANCOVA a MANOVA that includes a covariate

Matched samples/data synonym for paired samples or repeated measures

Mean a measure of position: sum of all values divided by the number of observations (synonym of arithmetic mean and average)

Mean square MS; used in ANOVA; the total sum of squares divided by the degrees of freedom (SS/df)

Measured variable a variable where the experimenter has to take a reading, e.g. height, weight, optical transmission

Measurement an observation or single item of data (datum)

Meta-analysis a method for combining the results of several tests of the same hypothesis. It can be used even when the original data is not available and different tests have been used

Median a measure of position: if all the data is put in rank order it is the value of the datum in the middle

MINITAB a statistical package

Mixed model (ANOVA) a test with both fixed and random grouping variables (factors)

Mode a crude measure of position: the most commonly occurring value

Model I ANOVA the basic ANOVA where the grouping variables (factors) are all fixed effects

Model II ANOVA an ANOVA where all the grouping variables (factors) are random effects

Model I regression the usual parametric regressions that assumes the 'cause' variable is measured without error

Model II regression rarely used version of regression that takes into account the fact that the 'cause' variable may be measured with error

Multifactorial design there are many variables that can be used to group the data

Multiple correlation difficult to interpret comparison of more than two variables

Multiple regression a test that establishes the best prediction of an 'effect' variable using all 'cause' variables simultaneously

Multivariate statistics tests which use more than one dependent variable

Negative binomial distribution a discrete probability distribution which is frequently invoked to describe contagious (clumped) distributions

Nested ANOVA synonym of hierarchical ANOVA. A test where at least one of the grouping variables is a sub-group of another (e.g. 'bunch' as a grouping variable within the grouping variable 'vine' in an experiment on grapes)

Nominal where the values in a data set cannot be put into any meaningful sequence only assigned to categories (e.g. blue and red)

Nonparametric test a test where few or no assumptions about the shape of a distribution are made

Normal distribution a unimodal probability distribution. This distribution is often assumed of the data in parametric statistics

Null hypothesis every hypothesis being tested must have a null hypothesis. E.g. if the hypothesis is that two groups have different mean heights then the null hypothesis must be that the two groups do not have different mean heights

Observation a single item of data (datum), a measurement

One-tailed test a test that assumes rejection of the null hypothesis can only come from a deviation in one direction rather than either. For example, the hypothesis that two groups are different is 'two-tailed' while the hypothesis that group 'A' is larger than group 'B' is 'one-tailed'. The effect of using a one-tailed test is to make statistics much less conservative for the same value of P

One-way ANOVA a parametric test of the null hypothesis that two or more groups come from the same population

Ordinal when values in a set of data can be placed in a meaningful order and ranked

Orthogonal literally means 'at right angles to'. In statistics it is used to indicate that two variables, factors or components are unrelated to each other

Outlier an extreme or aberrant observation lying well away from the rest of the data

***P* value** the probability of the significance statistic being that extreme or more if the null hypothesis is true. In biology the null hypothesis is usually rejected if the *P* value is <0.05.

Paired observations two observations taken from the same individual, perhaps in a 'before and after' design (synonym for repeated measures)

Paired samples sets of data comprising paired observations

Paired *t*-test to test the null hypothesis that two sets of observations on the same individuals (e.g. before and after) have the same distribution

Path analysis a complex data exploration system where there are several simultaneous and possibly interacting 'cause' and 'effect' relationships

Parametric test one where assumptions about the shape and spread of the data are made within the test

Partial correlation method of examining relationships between more than two variables that examines them pairwise while other variables are held constant

Pearson product–moment correlation coefficient the standard parametric correlation coefficient, *r*, measuring the association between two variables, that varies from 1 (perfect positive correlation), through 0 (no relationship) to –1 (perfect negative correlation).

Percentage when the relationship between two values is expressed as a single value (usually on a scale from 0–100)

Phi coefficient the result of a 2×2 contingency table

Pie chart a simple representation of frequencies of observations in different categories as sections (slices) of a circle. Particularly appropriate when the categories do not have a logical sequence

Platykurtic a distribution that has more observations in the 'shoulders' and fewer around the mean and in the tails than a normal distribution

Poisson distribution a discrete frequency distribution that results when events occur entirely at random

Polynomial regression a regression where the relationship between 'cause' and 'effect' is not assumed to be a straight line

Population the pool of possible individuals from which a sample is taken. Do not confuse with a biological population which will include additional individuals

Position the position of the sample is its midpoint which can be defined as a mean, median or mode

***Post hoc* test** meaning 'after this', there are several tests used after a one-way ANOVA to determine which groups are different from which

Power analysis a method based on the number of observations for determining the likelihood of detecting a statistically significant effect

Precision the range of possible values between which a particular observation may lie. The repeatability of a measurement (different to accuracy which is a measure of the closeness of a measurement to the true value)

Prediction interval in regression the range of *y* values that are expected for a given value of *x*. Usually given as a 95% prediction interval. This range does not fall within a zone running parallel to the best-fit line in linear regression but is smaller around the mid-range of *x*

Predictor in regression the 'cause' variable is often called the predictor. Always plotted on the *x*-axis

Principal axes a synonym for principal components

Principal component analysis a multivariate test which weights the variables to maximize the differences between individuals

Proportion when the relationship between two values is expressed on a scale from 0–1 (e.g. 30:60 becomes 0.5)

Proportional frequency a synonym for relative frequency

Pseudoreplication a problems in statistics when samples that are not independent are being treated as such

q–q plot a q–q or quantile–quantile plot is a graphical method for assessing whether a variable follows a normal distribution

Quadrat a square sampling area

Quadratic regression a regression where the relationship between the variables is assumed to be best described by a quadratic equation

Qualitative an observation that is assigned to a category that, although it may be coded as a number, has no numerical value (e.g. sex: coded 1 for female and 2 for male)

Quantitative an observation that has a meaningful numeric value. It can be either a direct observation or a count

Quartile when the data are ranked the quartiles are the values of the data points 25% and 75% down the list. They form the limits for the interquartile range

r the result of a Pearson product–moment correlation. If $r=0$ there is no correlation. If $r>0$ there is a positive relationship and if $r<0$ it is negative, 1 is perfect correlation, -1 is perfect negative correlation

r_s the statistic associated with the Spearman's rank correlation test

Random effect a term applied to factors in ANOVA that are not set by the experimenter (in contrast to a fixed effect)

Random sample a sample where each individual in the population has an equal chance of being measured or collected.

Randomized block design when sampling units are placed into groups (blocks) and the treatment applied to each sample is randomized within the block

Range a crude measure of dispersion: the distance from the lowest to highest value in a data set

Ratio when two values are expressed as a single number (e.g. $6:2$ becomes 3)

Raw data observations as they were originally recorded before any transformations or other processing is applied

Regression a description of the relationship between two variables where the value of one is determined by the value of the other (synonym of linear regression). More advanced regression can use several 'cause' variables

Related measures synonym for paired samples or repeated measures

Related samples synonym for paired samples or repeated measures

Relative frequency the proportion of observations having a particular value (or range of values). It is the frequency scaled to the sample size (e.g. 45 of 108 nests in a survey had four eggs, the relative frequency of four eggs is 45/108 or 0.417 or 41.7%)

Repeated measures two or more observations taken from the same individual at different times (if only two observations then this is a synonym for paired samples)

Repeated measures ANOVA analysis of variance carried out using repeat observations of the same individual. Time of observation (e.g. before and after) will be used as one of the factors in the ANOVA

Residuals the variation in the data left over after a statistical model has been accounted for (often regression or ANOVA). The model with the best fit has the smallest residual variation

Response in regression the 'effect' variable is often called this. Always plotted on the y-axis

Sample as all the individuals in a population may rarely be counted a portion of the population has to be taken, this is a sample

Sample size number of observations in a sample

Sample variance the variance of a single sample

Sampling unit the level at which an individual observation is made (e.g. a quadrat or a given size; a single leaf)

SAS a widely available and powerful statistical package

Scatter (plot) a graphical method for examining two (or possibly three) sets of data for possible relationships

Scheffé-Box test a test for heterogeneity of variance

Scheffé test one of many *post hoc* tests used in ANOVA to determine which groups are different from which

Scheirer–Ray–Hare test a nonparametric version of a two-way ANOVA, rarely supported in packages but quite easy to implement using the usual parametric ANOVA on ranked data and simple treatment of the resulting F value

Second-order interaction an interaction between three factors in ANOVA

Sign test a very conservative nonparametric test of a null hypothesis that there is no difference between two groups

Significance level the probability of achieving a significant result if the null hypothesis is true. In biology this is usually set at 0.05

Significant when the null hypothesis is rejected because the *P* value is less than 0.05 (the usual value in biology), 0.01 or any value set by the tester

Simple factorial (design) none of the grouping variables are sub-groups of any others, i.e. there is no nesting

Single-classification ANOVA synonym for one-way ANOVA (there is only one grouping variable)

Skew(ness) a measure of symmetry (sometimes called g_1). Positive skew indicates that there are more values in the right tail of a distribution than would be expected in a normal distribution. Negative skew indicates the left tail.

Skewed distribution a distribution that has a value of skewness other than zero

Slope a number (usually *b* or β) denoting how a trend line deviates from zero (a slope of 0)

SNK test Student–Newman–Keuls test, a common *post hoc* test

Spearman rank-order correlation a nonparametric measure of correlation

Spearman's rank correlation alternative name for Spearman rank-order correlation

Split-plot design an experimental design technique used to analyse two factors when there is only one 'plot' for each level of one of the factors

Spread the way in which the data are distributed, often measured by standard deviation (synonym of dispersion)

SPSS a widely available statistical package

Standard deviation a measure of spread: sensitive to shape of distribution

Standard error a measure of spread: the standard deviation of the values of a set of means taken from a data set. Sensitive to sample size

Stem and leaf chart a method of displaying the data commonly used by computers before graphical output was possible. The 'stem' would be represented by a series of rows and leaf by columns starting from the left. Each observation would be assigned to a position on the stem based on its value

Stepwise regression a regression analysis where the best method for predicting the 'effect' from several 'cause' variables is sought

Stratified random sample a method of collecting a sample that takes into account a feature of the collecting area

Student pseudonym used by the statistician William Gossett

Student's *t*-test synonym for independent samples *t*-test

Student–Newman–Keuls (SNK) test a frequently used *post hoc* test, used after a one-way ANOVA to determine which groups are different from which

Summary statistic anything that condenses the information about a variable, such as a mean or standard deviation

Symmetry a data set with symmetry has the same shape either side of the mean

***t*-distribution** a family of distributions widely used in statistics that is derived from the distribution of sample means with respect to the true mean of a population

***t*-test** to test the null hypothesis that two groups come from the same distribution (synonym for Student's *t*-test, independent samples *t*-test)

Tally when observations are assigned to categories and marked as ticks in a table

Three-way… an experiment involving three independent factors. See 'two-way…' and extrapolate

Time series a set of data points taken at different points in time

Transformation a mathematical conversion that is applied to every observation in a data set. Usually used to make a distribution conform to a normal distribution

Treatment a level (usually denoted by an integer) of an independent variable (i.e. set or defined by the experimenter)

Tukey test one of many *post hoc* tests used in ANOVA to determine which groups are different from which

Tukey–Kramer method synonym for Tukey test. A *post hoc* test, used after a one-way

ANOVA to determine which groups are different from which

Two-tailed test applies to most statistical tests and implies that the null hypothesis can be rejected by deviations either up or down. For example, if the null hypothesis that two groups of bats use the same frequency for echo location is rejected then group 'A' may use either a significantly higher or significantly lower frequency than group 'B'. If the standard $P = 0.05$ level is used then it implies a $P = 0.025$ region in each tail

Two-way ANOVA an ANOVA test where there are two ways of grouping the data

Two-way interaction in ANOVA this a measure of whether two grouping variables have an additive (no interaction) effect or not (interaction)

Type I error a truly significant result is deemed nonsignificant by a test

Type II error a truly nonsignificant result is deemed significant by a test

Unbalanced when there are different numbers of observations in different factor combinations. Severely unbalanced designs (i.e. where some of the factor combinations have no observations) should be avoided

Uniform distribution a 'flat' distribution where the chance of any value occurring is approximately equal, may often be transformed, using the arcsin transform, to an approximately normal distribution

Unimodal a frequency distribution with a single peak at the mode

Univariate statistics statistical tests using only one dependent variable

Unpaired data synonym for independent data, stressing that sets of data are not paired

Value a single piece of data (datum)

Variable anything that varies between individuals (e.g. 'sex', 'weight' or 'aggressiveness'). The term variate is actually correct, but variable is now the widely used term for the observed data set.

Variance the sum of squared deviations of observations from the mean: a measure of spread of the data. Very important in the mechanics of statistics but not very useful as a descriptive statistic

Variance/mean ratio [v/m or s^2/m] a commonly quoted statistic useful for describing whether a set of observations fits a Poisson distribution ($v/m = 1$), is more clumped ($v/m > 1$) or is more ordered ($v/m < 1$)

Variate the correct term for variable. Still retained for terms such as canonical variate analysis or univariate statistics

Weibull distribution a family of continuous distributions

Welch's approximate t-test a version of the Student's t-test that can be used when the variances of the two samples are known to be unequal

Welsch step-up procedure a *post hoc* test, used after a one-way ANOVA to determine which groups are different from which; requires equal sample sizes

Wilcoxon–Mann–Whitney test synonym for the Mann–Whitney U test. A nonparametric test of a null hypothesis that two groups come from the same distribution

Wilcoxon's rank sum W test synonym for the Mann–Whitney U test. A nonparametric test of a null hypothesis that two groups come from the same distribution

Wilcoxon's signed rank test a nonparametric test of a null hypothesis that there is no difference between two related groups. Equivalent to a paired t-test

Williams' correction a method of correcting bias in various contingency tests such as the G-test

Winsorize a method used to reduce the effect of outlying observations by replacing them with the next value towards the median

Within sometimes used as shorthand for 'within sample variance' or 'within group variance'

Within sample variance in ANOVA a measurement of the amount of variation within a sample (see between sample variance)

x-axis the horizontal axis of a graph or chart (abscissa)

Yates' correction sometimes called the 'continuity correction'. A method to make the results of a 2×2 chi-square test more conservative

y-axis the vertical axis of a graph or chart

z-axis the axis that 'goes into the paper or screen' on a three-dimensional graph or chart

z-distribution occasionally used as a synonym for the normal distribution

Glossary **z-test** a test used to compare two distributions, or more usually to compare a sample with a larger population where the mean and standard deviation are known

Bibliography and short reviews of selected texts

Atkinson, A.C. & Doney, A.N. (1992) *Optimum Experimental Design.* 328pp. Oxford University Press, Oxford.

An advanced statistics book covering the theory of model fitting. Only recommended for those with a high level of mathematical knowledge.

Bailey, N.T.J. (1995) *Statistical Methods in Biology*, 3rd edition. 255pp. Cambridge University Press, Cambridge.

Now a rather venerable book (first edition was 1959), this has been used by innumerable statistics courses in biology. It has a very traditional feel and takes the conventional path from data collection through to fairly complex analyses. Good points: summary of statistical formulae excellent for the statistics initiate; very good considerations of the assumption of the tests. Bad points: nonparametrics added on as something of an afterthought; too many formulae.

Barnard, C., Gilbert, F. & McGregor, P. (1993) *Asking Questions in Biology. Design, Analysis & Presentation in Practical Work.* 157pp. Longman Scientific & Technical, Harlow.

This is an excellent companion for a first term's experimental biology. Good points: it covers the basic concepts of experimental design in a superb way; has a simple key to tests and offers good advice on data presentation. Bad points: not many statistical tests included; no advice on the use of computer packages.

Bland, M. (1995) *An Introduction to Medical Statistics*, 2nd edition. 396pp. Oxford University Press, Oxford.

Although medically biased this book is an excellent combination of the traditional statistics book and a modern outlook and style. Good points: covers a very wide range of statistical tests, data presentation and manipulation in a very detailed yet approachable way; some concepts placed in thoughtful appendices; interesting exercises with solutions; multiple choice questions; nonparametric techniques properly integrated. Bad points: assumes that all the statistical work will be done using pen and paper.

Bryman, A. & Cramer D. (1997) *Quantitative Data Analysis with SPSS for Windows: a Guide for Social Scientists.* 313pp. Routledge, London.

This book attempts to do much the same as this volume —minimizing the statistical theory and maximizing the understanding of the correct use of the statistical package. Good points: lots of screen shots or SPSS; uses SPSS for Windows: exercises with answers. Bad points: examples are all for sociology with a little psychology; restricted to SPSS; no coverage of experimental design.

Campbell, R.C. (1989) *Statistics for Biologists*, 3rd edition. 446pp. Cambridge University Press, Cambridge.

A book that covers a lot of ground in an approachable way. Good points: considered from the problem side using nonparametric statistics sensibly and frequently; distribution and experimental design covered clearly; many exercises; statistical packages used. Bad points: statistical packages rather dated; more consideration of the theoretical background than is necessary for biologists.

Chatfield, C. (1995) *Problem Solving. A Statistician's Guide*, 2nd edition. 317pp. Chapman & Hall, London.

Although not biological, this engaging and well-written book is worth a look at. It highlights a large number of possible difficulties in statistical analysis and offers advice on how things should be done better. There is also an appendix that provides a digest of statistical techniques that would be useful as a quick reference for anyone who already knows how the statistic works.

Clarke, G.M. (1994) *Statistics and Experimental Design. An Introduction for Biologists and Biochemists*, 3rd edition. 208pp. Edward Arnold, London.

This book has been revised somewhat over earlier editions but still retains much of its traditional feel. Good points: there are some good tips here; a large number of exercises; ANOVA considered from the experimental design side. Bad points: more mathematics than is necessary in the text; nonparametric statistics are treated as second-class citizens; no consideration of 'nested' ANOVA.

Fowler, J. & Cohen, L. (1990) *Practical Statistics for Field Biology*. 227pp. John Wiley & Sons, Chichester.
A very well presented 'traditional' introductory statistics book. Covers the basics to simple ANOVA and regression. Good points: careful coverage of distributions, sampling, variation. Bad points: assumes all the statistics are carried out by hand.

Fowler, J. & Cohen, L. (1995) *Statistics for Ornithologists, BTO Guide 22*, 2nd edition. 150pp. British Trust for Ornithology, Thetford.
A short and approachable traditional statistics that covers the basics up to two-way ANOVA using examples that are nearly all based on birds. Good points: useful appendices; careful explanations of types of data; short enough not to be intimidating. Bad points: assumes all the statistics are carried out by hand.

Gilbert, N. (1973) *Biometrical Interpretation*. 125pp. Oxford University Press, Oxford.
A rather old text but worth considering because of its rather unorthodox approach to the subject. Good points: considers the theory and meaning of statistical tests in a biological context; excellent text to accompany a traditional statistics book. Bad points: no explanation of the tests themselves.

Glantz, S.A. (1997) *Primer of Biostatistics*, 4th edition. 472pp. McGraw-Hill, New York.
This is one of the most approachable of the traditional statistics books. Good points: clear diagrams; lots of examples; exercises with answers; careful explanation of major concepts. Bad points: examples are medical only; lots of equations; nonparametric tests treated separately; no two-way ANOVA.

Hays, W.L. (1994) *Statistics*, 5th edition. 1112pp. Harcourt Brace, Fort Worth.
This is a huge text starting from the very basics and ending with some really quite complicated stuff. There are extensive exercises after each chapter and long appendices if you require more maths. This book does not use biological examples and is aimed at a general audience. It is also rather detailed and should only be recommended to those seeking to invest a lot of time in understanding how statistics actually function.

Heath, D. (1995) *An Introduction to Experimental Design and Statistics for Biology*. 372pp. UCL Press, London.
A 'traditional' but approachable statistics book with lots of biological examples and a strong emphasis on experimental design and sampling strategies. Comprehensive coverage of parametric statistics up to ANOVA with nonparametric equivalents included. Good points: calculations are put into text boxes; a key to statistical tests; good glossary. Bad points: all calculations done by hand; key not extensive enough.

Levine, G. (1991) *A Guide to SPSS for Analysis of Variance*. 151pp. Lawrence Erlbaum Associates, Hillsdale, NJ.
A short and detailed but restricted book covering a variety of analysis of variance designs. Unfortunately the version of SPSS assumed is the command line version.

Manly, B.F.J. (1994) *Multivariate Statistical Methods, A Primer*, 2nd edition. 215pp. Chapman & Hall, London.
This book is short, not biology-based and rather mathematical but the text is very clearly written. It is an excellent introduction to the way that some of the more common multivariate analyses actually work.

Maxwell, S.E. & Delaney, H.D. (1990) *Designing Experiments and Analyzing Data. A Model Comparison Perspective*. 902pp. Wadsworth, Belmont CA.
An advanced book dealing with the theory of experimental design and then going onto some advanced statistics. The authors assume you will be using a computer but feel that some hand calculation of the statistics will help with understanding. This book should be used as a reference work and is particularly useful for the more complicated ANOVA designs. Good points: in-depth coverage of analysis of covariance; considers experimental design as the key to good science; many exercises. Bad points: many equations, matrices, etc.; solutions to many of the exercises only available separately.

Bibliography

Mead, R. (1988) *The Design of Experiments. Statistical Principles for Practical Applications.* 620pp. Cambridge University Press, Cambridge.
An advanced book that deals with the theory of experimental design along with a lot of the mathematics behind experimental design concepts such as randomization, blocking and replication. Only recommended for those with solid mathematics grounding. Good points: wide coverage of experimental design theory. Bad points: too many equations.

Mead, R. & Curnow, R.N. (1983) *Statistical Methods in Agriculture and Experimental Biology.* 335pp. Chapman & Hall, London.
A 'traditional' style statistics book covering all the basic techniques of parametric analysis with a good emphasis on appropriate sampling and experimental design. Good points: very good explanation of the assumptions of tests. Bad points: very few nonparametric tests.

Miller, R.G., Efron, B., Brown, B.W. & Moses, L.E. (eds) (1980) *Biostatistics Casebook.* 238pp. John Wiley & Sons, New York.
An old book with a medical bias. Interesting as it deals with specific examples in some detail. A mixture of advanced and basic techniques.

Montgomery, D.C. (1991) *Design and Analysis of Experiments*, 3rd edition. 649pp. John Wiley & Sons, New York.
This is an introductory book for those studying statistics to degree level. Consequently it does cover the mathematics and the theory of statistics in some detail. However, it does include information from the very basics of experimental design. Good points: complete coverage up to multifactorial designs; includes sample problems; some consideration of nonparametric alternatives. Bad points: many equations, matrices, etc.; no solutions to the problems.

Parker, R.E. (1979) *Introductory Statistics for Biologists*, 2nd edition. Studies in Biology; no. 43. 122pp. Edward Arnold, London.
A small but dense, traditional text that has been recommended to many undergraduate biologists. Good points: a huge amount of information packed into a very small book; statistical tables included; exercises with solutions. Bad points: little consideration of nonparametric statistics; useful only if accompanied by a taught course.

Sanders, D.H. (1995) *Statistics: A First Course*, 5th edition. 650pp. McGraw-Hill, New York.
Starting from the very beginning, this book leads the reader through a wealth of examples and exercises to some basic ANOVA and regression. Good points: easy to understand; clearly illustrated with graphs, etc.; no unnecessary equations; in 5th edition. Bad points: only uses 'command line' MINITAB as a computer package; no biological orientation; nonparametric tests relegated to the last chapter.

Scheiner, S.M. & Gurevitch, J. eds. (1993) *Design and Analysis of Ecological Experiments.* 443pp. Chapman & Hall, New York.
An excellent book divided into discrete chapters each covering a specific ecological example and statistical technique. Good points: very good, detailed description of often complex statistics (e.g. path analysis and multiple regression); assumes use of computers. Bad points: only uses the SAS statistical package; coverage far from complete.

Siegel, S. & Castellan, N.J. (1988) *Non-parametric Statistics for the Behavioural Sciences*, 2nd edition. 399pp. McGraw-Hill, New York.
This edition is a major revision of the 1956 first edition by Siegel. The title shows where nonparametric statistics first made their mark in biology. Good points: very wide range of nonparametric statistics covered; compares power with parametric equivalents; summary inside the covers; lots of behavioural examples. Bad points: often gives several methods of achieving the same thing; mathematics in the text.

Sokal, R.R. & Rohlf, F.J. (1995) *Biometry*, 3rd edition. 887pp. Freeman, New York.
This large tome is almost a statistics 'bible' for many research biologists. The latest edition has made some concessions to the advent of the computer. Good points: covers a huge range of statistics; calculation methods are kept in boxes away from the text; summary inside the covers; recent innovations included. Bad points: too large and daunting for the faint-hearted; nonparametric statistics not fully integrated; statistical tables in a separate volume.

Bibliography

Sprent, P. (1993) *Applied Nonparametric Statistical Methods*, 2nd edition. 342pp. Chapman & Hall, London.

An excellent book for those wishing to understand how nonparametric statistics actually work. The book is surprisingly easy to read given the subject matter and the coverage of methods is very wide. Good points: coverage of many tests not available in statistical packages; consideration of the advantages of nonparametric methods; simple examples. Bad points: examples not biological; not suitable for a statistical novice; statistical packages referred to are specialist.

Steel, R.G.D. & Torrie, J.H. (1980) *Principles and Procedures of Statistics. A Biometrical Approach*, 2nd (international) edition. 632pp. McGraw-Hill, New York.

Revised after twenty years and still has a traditional feel. Good points: covers everything in some detail; worked exercises in each chapter; very good if you want to know how a test works; covers experimental design. Bad points: in at the deep end for a beginner; lots of mathematics in the text; nonparametric statistics segregated.

Underwood, A.J. (1997) *Experiments in Ecology. Their Logical Design and Interpretation Using Analysis of Variance*. 504pp. Cambridge University Press, Cambridge.

This book goes into a lot of detail describing sampling, variation and experimental design. It certainly achieves its aim to encourage the correct design of experiments for analysis of variance techniques. Too advanced for beginners and restricted in scope but very useful for anyone designing large or complicated experiments involving factorial designs.

Watt, T.A. (1993) *Introductory Statistics for Biology Students*. 185pp. Chapman & Hall, London.

A book that starts from the beginning with the collection of data. Covers sampling methods, basic and more advanced experimental design and most basic statistics. Good points: easy to read; starts assuming no knowledge; aimed at computer users; advice on incorporation of statistics into reports; no equations. Bad points: only covers basic statistics; uses command line MINITAB as the only statistical package.

Wright, D.B. (1997) *Understanding Statistics: An Introduction for the Social Sciences*. 228pp. Sage Publications, London.

I have included this book despite the title because I feel that biologists would find it useful. It is a traditional book in that it covers the mathematics behind the statistics to some extent although it makes extensive use of computer-based examples. Good points: assumes no knowledge; uses SPSS; exercises; covers everything from means to quite advanced techniques. Bad points: only uses command line SPSS; examples are not biological; not all the exercises have answers.

Index

Note: Page numbers in *italics* refer to Figures

abscissa 43
accuracy 32, 33
aggregated population or distribution 34, 35
analysis of covariance (ANCOVA) 142, 171,
 172–5
 MINITAB 174–5
 SPSS 173–4
analysis of variance (ANOVA) 25
 hierarchical 142
 multiway 142
 nested 142–146
 one-way for independent samples 99–105
  Excel 104–5
  MINITAB 103–4
  SPSS 100–3
 one-way for unpaired data 85–92
  Excel 91–2
  MINITAB 90–1
  SPSS 86–90
 post hoc testing 108–10
  MINITAB 109–10
  SPSS 108–9
 repeated measures 98
 three-way 137–141
 two-level nested 143–6
  MINITAB 146
  SPSS 144–6
 two-way (with replication) 123–31
  Excel 129–31
  MINITAB 127–9
  SPSS 125–7
 two-way (without replication) 114–20
  Excel 119–20
  MINITAB 118–19
  SPSS 115–18
Anderson–Darling (A–D) test 37, 40, 61, 72
 MINITAB 72
ANCOVA *see* analysis of covariance
ANOVA *see* analysis of variance
arcsin square root transformation (for
 percentage or proportion data) 40
arithmetic mean 46
association, tests of 170–1
 correlation 170–1
 Kendall's partial rank-order correlation 171
 partial correlation 171
 associations
 comparing two variables 53–6
 comparing more than two variables 57–9
asymmetric distribution 48
attributes 33
average *see* mean

bar chart (for discrete data) 43
bias 3
bimodal distribution 40, 47, *47*
binomial distribution 34–7

blocks 28
box and whisker (box) plot 42, 43, *44*, 47, 53

canonical variate analysis (CVA) 182
categorical observations 2
categorical variables 33
chi-square distribution 41
chi-square test 36, 37, 40
 as test of association 147–53
  Excel 152–3
  MINITAB 151–2
  SPSS 149–51
 as test of difference 61–70
  Excel 68–70
  MINITAB 65–7
  SPSS 62–5
 goodness of fit *see* chi-square test of
  difference 61–70
choosing a test 7–21
cluster analysis 185–6
 MINITAB 186
 SPSS 186
coefficient of variation (CV) 49
computed variable 33
confidence intervals (CI) 41, 49, 53, 57, 58,
 164
confidence limits 49
continuous distribution 37–40
continuous observations 2
continuous variable 32
control 26–8
 experimental 27
 procedural 26–7
 statistical 27–8
 temporal 27
correlation 5, 170–1
 Kendall's partial rank-order correlation
  171
 Kendall's rank-order correlation 160–1
  SPSS 161
 Partial correlation 171
 Pearson's product-moment correlation
  154–7, 158, 159–60
  Excel 157
  MINITAB 156–7
  SPSS 155–6
 Spearman's rank-order correlation 154,
  158–60
  Excel 159–60
  MINITAB 159
  SPSS 158
Cramér coefficient of association 153

DECORANA 187
degrees of freedom 41
derived variable 33
descriptive statistics 30, 46–9
 MINITAB 51
 SPSS 50

statistics of distribution, dispersion or spread 48–9
statistics of location 46
discontinuous variable 32
discrete distributions 34–7
discrete observations 2
discrete variable 32
discrete variation 32
discriminant function analysis 182–5
  MINITAB 184–5
  SPSS 182–4
distribution-free tests 31
distributions 34–40
  describing 38–40
dummy data, collection of 1

error bars 53
errors 22–3
Excel 52 (see tests)
experimental control 27
experimental design 26–9
experiments 4
exponential distribution 41–2

factor 4
factor analysis 177–82
factor levels 4
factorial test 5
Fisher's LSD test 108
fixed factors 142–143
fitted lines 56–7
flat distribution 37
Friedman test (for repeated measures) 95–8
  MINITAB 97–8
  SPSS 96–7
Friedman test (no replication) 111–14
  MINITAB 113–14
  SPSS 112–13

Gaussian distribution 37
geometric mean 46
G-test 36, 61

harmonic mean 46
hierarchical designs 143–6
histogram for continuous data 43, 44–5, 45
hypergeometric distribution 37
hypothesis
  definition 22, 30
  testing 2, 22

independent samples 98–110
interaction 121–3
  Excel 122–3
  MINITAB 122
  SPSS 121–2
inter-quartile range 42, 48

Kendall's partial rank-order correlation 171
Kendall's rank-order correlation 160–1
  SPSS 161
Kendall's robust line-fit method 169
Kendall's Tau (*T*) 160
key to tests 7–21
Kolmogorov–Smirnov (K–S) test 37, 40, 61, 62, 70–2

one-sample test 70
SPSS 71–2
two-sample test 70
Kruskal–Wallis test 106–8
  MINITAB 107–8
  SPSS 106–7
kurtosis 40, 49

Latin square 28
least significant difference (LSD) test 108
leptokurtic distribution 40
Levene test 73, 82, 83
linear regression 163
  Excel 168–9
  MINITAB 167–8
  SPSS 165–7
log transformation 40
logistic regression 169

Mann–Whitney *U* 70, 92–5
  MINITAB 94–5
  SPSS 93–4
matched data 72–80
matched samples 95–8
mean 4, 30, 37, 41
  arithmetic 46
  geometric 46
  harmonic 46
mean square *see* variance
median 4, 30, 42, 47
meristic variable 32
MINITAB 51–2, *52* (see tests)
mixed model 143
mode 47–8
model I linear regression *see* standard linear regression
model II regression 170
multifactorial testing 137–42
multimodal distributions 47, *47*
multiple fitted lines 59
multiple regression 171, 175
multiple scatterplots 54–6
multivariate analysis of covariance (MANCOVA) 185
multivariate analysis of variance (MANOVA) 185
multivariate technique 6
multiway ANOVA 142

negative binomial distribution 36–7
nested factors 142–3
nested/hierarchical designs 143–6
nominal variables 33
nonindependent factors 142
non-parametric distribution 42
non-parametric statistics 31
normal distribution 37–40, *38*
  convergence of a Poisson distribution to 38
  perfect 37, 40
  sampling distributions and 'central limit theorem' 38
  standardized 37–8
null hypothesis 22, 30, 37
  definition 2

observations 2

one-way ANOVA for independent samples
  99–105
  Excel 104–5
  MINITAB 103–4
  SPSS 100–3
one-way ANOVA for unpaired data 85–92
  Excel 91–2
  MINITAB 90–1
  SPSS 86–90
one-way ANOVA, *post hoc* testing 108–10
  MINITAB 109–10
  SPSS 108–9
ordinate 43

paired data 72–80
paired *t*-test 73–5
  Excel 75
  MINITAB 74–5
  SPSS 73–4
parametric statistics 31
partial correlation 171
path analysis 176
Pearson's product-moment correlation
  154–7, 158, 159–60
  Excel 157
  MINITAB 156–7
  SPSS 155–6
phi coefficient of association 153
pie chart (for categorical data or attribute
  data) 45, *46*
platykurtic distribution 40
Poisson distribution 34, 37, 62
polynomial regression 170
post-hoc tests 108–10
  MINITAB 109–10
  SPSS 108–9
precision 32
principal component analysis (PCA) 177–82
  MINITAB 181–2
  SPSS 180–1
probits 40
procedural control 26–7
P-values 3, 22, 23

quadrat-variance methods 24
quadratic regression 170
quadrats 25
quartiles 42, 48

random (or fixed) factors 143
random walks 25
range 48
ranked tests 31
ranked variable 33
ranking 42
ranking tests 31
rectangular (flat or uniform) distribution 37
regression 5, 161–2, 172
  confidence intervals and 57, *58*
  multiple 171,175
  polynomial 170
  quadratic 170
related data 72–80
related samples 95–8
repeated measures 95–8
repeated measures ANOVA 98

sampling 3, 23–5
  choice of sample unit 24
  number of sample units 24
  positioning of sample units 24–5
  random 24–5
  size of sample unit 24
  timing 25
scatterplots 54, *55*
  multiple 54–6
Scheirer–Ray–Hare test 131–6
  Excel 134–6
  MINITAB 133–4
  SPSS 131–3
Sign test 77–80
  Excel 79–80
  MINITAB 78–9
  SPSS 78
skewness 38–40, 49
Spearman's rank-order correlation 154,
  158–60
  Excel 159–60
  MINITAB 159
  SPSS 158
SPSS 50–1, *50*, *54* (see tests)
square-root transformation 40
standard deviation 4, 37, 41, 48, 49
standard error (SE) 38, 41, 48–9
standard linear regression 163
  confidence intervals 164
  cf correlation 164
  Excel 168–9
  MINITAB 167–8
  prediction 163
  prediction intervals 164
  residuals 164
  SPSS 165–7
statistical control 27–8
statistics
  definition 4, 30
  descriptive 4
  tests for data investigation 5–6
  tests of difference 4–5
  tests of relationships 5
  types of 30–1
stepwise regression 175–6
stratified random assignment 29
stratified random sample 25
Student Newman Keuls (SNK) *post hoc* test
  108
successful data analysis, eight steps to 1, 7
surfaces 59–60

t-distribution 40–1
temporal control 27
three-dimensional graphs, difficulties with 59
three-dimensional scatterplots 57–9
three-way ANOVA (with replication) 137–41
  MINITAB 140–1
  SPSS 139–40
three-way ANOVA (without replication) 137
transformations 40
treatments 4
t-test 61, 70, 80–5
  Excel 84–5
  MINITAB 83–4
  SPSS 81–3

TWINSPAN 187
two-level nested design ANOVA 143–6
  MINITAB 146
  SPSS 144–6
two-way ANOVA (with replication) 123–31
  Excel 129–31
  MINITAB 127–9
  SPSS 125–7
two-way ANOVA (without replication) 114–20
  Excel 119–20
  MINITAB 118–19
  SPSS 115–18
Type I errors 22–3
Type II errors 22–3

uniform distribution 37
uniform population 34, 35
unimodal distributions 47, *47*, 48
unpaired data 80–95

variable 31
  continuous vs discrete 32
  definition 3, 31, 48
  selecting best 31
  types 31–3
variance 48

Wilcoxon rank sum *W* test 70, 92–5
Wilcoxon–Mann–Whitney test 70, 92–5
Wilcoxon's signed ranks test 75–7
  MINITAB 76–7
  SPSS 76
William's correction 61

x-axis 43–44, *46*, 56
x-bar (mean) 46

y-axis 43–44, 56